Ka-52 Hokum

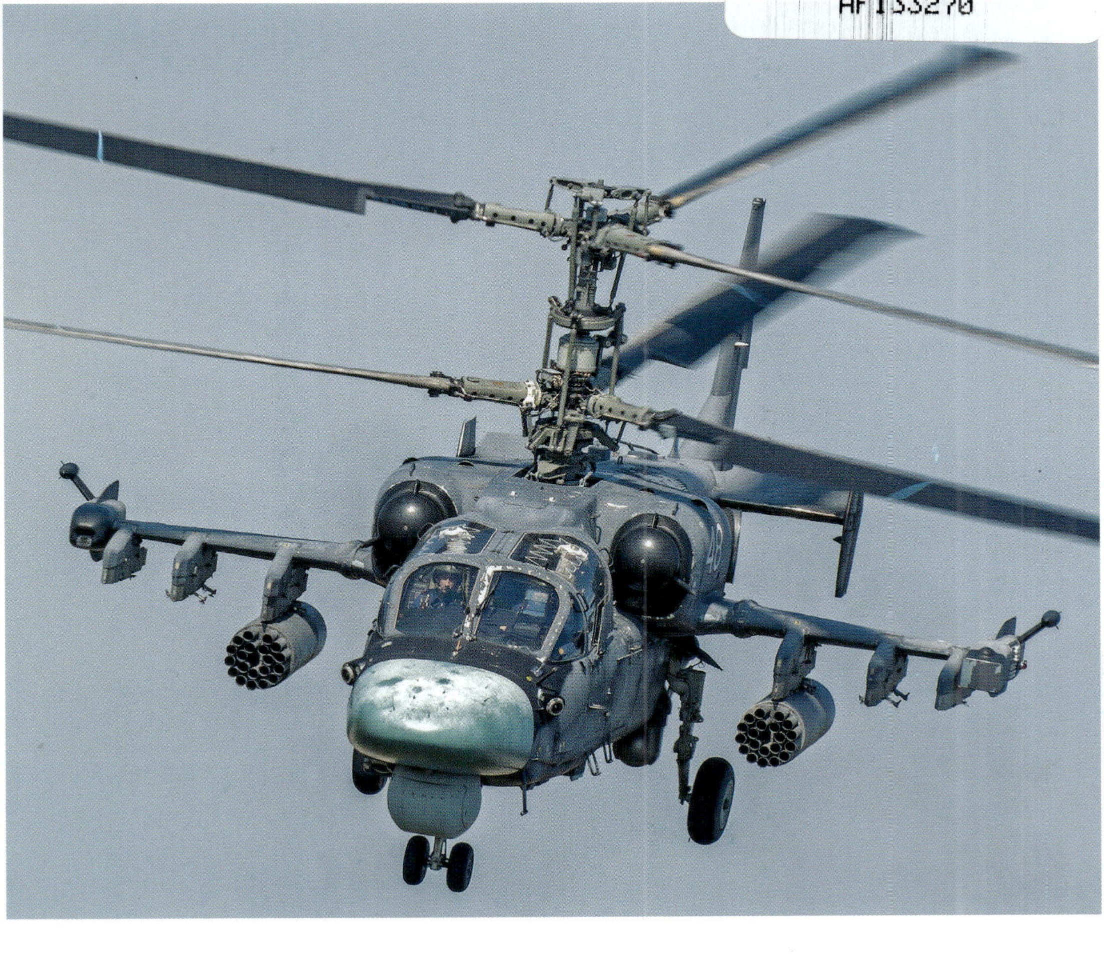

PIOTR BUTOWSKI

KEY Books

MODERN MILITARY AIRCRAFT SERIES, VOLUME 8

Front cover image: The Ka-52 *Hokum* stands out from other combat helicopters in the world thanks to its two co-axial rotors, as well as the positioning of the pilots side by side. In the wide front fuselage, there is room for a large radar. The wing does not have an aerodynamic function but is used to carry weapons and equipment. (Piotr Butowski)

Back cover image: Ka-52 '90' RF-13429 of 55th Independent Helicopter Regiment in Korenovsk. (Piotr Butowski)

Title page image: Ka-52 '48' RF-91332 belongs to Ostrov-based 15th Army Aviation Brigade of the Russian Aerospace Forces, which have been taking part in the invasion of Ukraine that started on 24 February 2022. (Piotr Butowski)

Contents page image: Ka-52 '84' RF-13423 of 55th Independent Helicopter Regiment in Korenovsk. (Piotr Butowski)

Published by Key Books
An imprint of Key Publishing Ltd
PO Box 100
Stamford
Lincs PE19 1XQ

www.keypublishing.com

The rights of Piotr Butowski to be identified as the author of this book has been asserted in accordance with the Copyright, Designs and Patents Act 1988 Sections 77 and 78.

Copyright © Piotr Butowski, 2022

ISBN 978 1 80282 269 4

All rights reserved. Reproduction in whole or in part in any form whatsoever or by any means is strictly prohibited without the prior permission of the Publisher.

Typeset by SJmagic DESIGN SERVICES, India.

Contents

Chapter 1 Hokum's prolonged start 4
Chapter 2 Ka-50 in production and service 17
Chapter 3 For export: the first attempt 31
Chapter 4 Ka-52 two-seater: the beginning 39
Chapter 5 Ka-52 in production and service 45
Chapter 6 Ka-52 in detail 58
Chapter 7 Ka-52E for export 74
Chapter 8 Ka-52K Katran ship-based helicopter 78
Chapter 9 Modernised Ka-52M 85
Chapter 10 What is next? 94

Chapter 1
Hokum's prolonged start

In the second half of the 1960s, there was an avalanche-like expansion of American helicopters used in Vietnam; their number there increased from 400 in 1965 to 4,000 in 1970. In the USSR, this development was observed, and lessons were learned. On 29 March 1967, Mikhail Mil's design bureau was charged with the development of a combat helicopter concept.

At the time, the Soviet combat helicopter design was different to that in the West: in addition to weapons, it was also supposed to carry a squad of soldiers. This idea arose from the enthusiasm of Soviet military commanders after the introduction of the BMP-1 infantry fighting vehicle (IFV) to the armament of the Soviet Army in 1966. The armoured vehicle carried eight soldiers and was armed with a 73mm 2A28 gun and Malyutka anti-tank guided missiles. Its use opened up new tactical possibilities

Left: The Ka-25F army helicopter design used the engines, main gearbox and rotors of a naval Ka-25 *Hormone*. It lost in the competition to the Mi-24 *Hind* by Mikhail Mil. (Piotr Butowski)

Below left: There was space for 12 troops in the Ka-25F's hold. A movable cannon was attached under the fuselage. (Piotr Butowski)

Below right: Kamov's deputy general designer, Veniamin Kasyanikov, who introduced the author to the history of combat helicopter designs that lead to the Ka-50 and Ka-52, 1993. (Piotr Butowski)

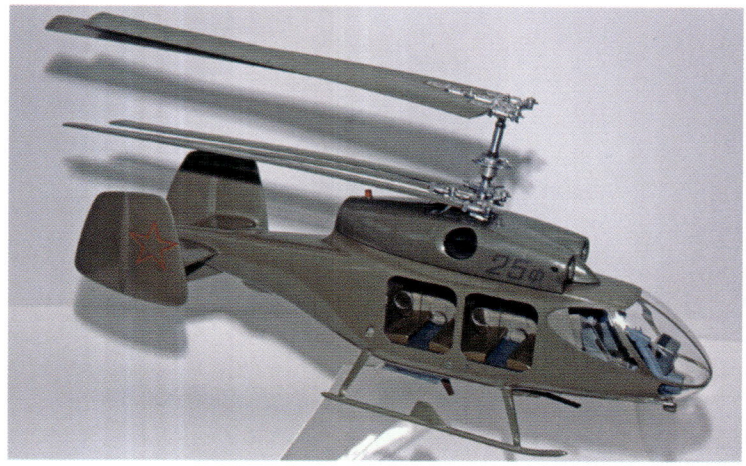

Above left: The V-50 of 1969 was a tandem twin-rotor combat and transport helicopter expected to reach 216kts (400km/h). (Piotr Butowski)

Above right: The V-80 design from 1971 was the first of Kamov's purely combat army helicopters, without transport capacity. The crew consisted of a pilot and a gunner sitting in tandem. Note the retractable undercarriage and fixed gun. (Piotr Butowski)

for the land forces. The idea to go a step further was proposed and the Soviet helicopter designers were ordered to develop a 'flying IFV'.

Originally, the commission was only given to Mikhail Mil, as Nikolai Kamov had previously only made seaborne helicopters; he worked only with the navy and was not considered as a contender by the Soviet Army Aviation. However, when Nikolai Kamov found out about the order for an army combat helicopter, he put himself forward for the project.

Kamov developed the Ka-25F (frontovoi, tactical) project, emphasising its low cost, obtained thanks to the use of elements of its latest Ka-25 naval helicopter, which, from April 1965, was mass-produced at the Ulan-Ude plant. The Ka-25 had a unique configuration in that the engines, main gearbox and rotors were together in an easy-to-remove one-piece module. Kamov proposed to use this module in the new army helicopter and add a new fuselage to it. In the cockpit, the pilot and the weapons operator sat side by side, with room in the hold for 12 troops. In the combat variant, instead of soldiers, the helicopter could take anti-tank guided missiles (ATGMs) on external pylons. There was a 23mm cannon on a movable turret under the nose. While working on the Ka-25F, the Kamov team experimented with a Ka-25, from which the radar and anti-submarine warfare equipment were removed, and 57mm unguided rocket launchers were installed. The designers planned the skid chassis for the Ka-25F to be more durable than the wheeled landing gear. Later, they considered this a mistake, as the use of a skid undercarriage is only suitable for very light helicopters.

The Ka-25F was supposed to be a small helicopter; according to the design, it had a weight of 17,637lb (8,000kg) and two GTD-3F turboshaft engines of 671kW (900shp) each, made by the Valentin Glushenkov design bureau in Omsk. It was planned to increase the power to 932kW (1,250shp). However, during the course of the programme, the military's requirements grew, and it was no longer possible to meet them within the size and weight limitations of the Ka-25 helicopter. For example, the military demanded armour for the landing cabin and pilot's cockpit, which was not in the original specification. The GTD-3 engines could no longer cope with such a load. Meanwhile, Mikhail Mil's team did not limit themselves to existing solutions and prepared their '240' project – the Mi-24, a completely new helicopter powered by modern and powerful Klimov TV2-117 engines (1,119kW [1,500shp] each).

Thus, the Ka-25F lost to the Mi-24 in the design competition. On 6 May 1968, a joint resolution of the Central Committee of the Communist Party of the Soviet Union (CPSU) and the USSR

Above left: The V-100 compound helicopter design of the mid-1970s had two side-by-side rotors and a small pushing propeller at the fuselage rear that allowed the helicopter to reach a speed of 216kts (400km/h). (Piotr Butowski)

Above right: The V-100 had a powerful armament, including these Kh-25 anti-ship or Kh-27 anti-radar missiles, two cannons in side pods and six underwing pylons. (Piotr Butowski)

government ordered the new combat helicopter from the Mikhail Mil design bureau. As the 'flying infantry fighting vehicle' was a high-priority task, on 19 September 1969, the '240' prototype performed its first flight, and, in November 1970, the production plant in Arsenyev produced the first Mi-24. Today, more than 3,700 Mi-24 (NATO reporting name: *Hind*) helicopters have been made and, in the Mi-35M form, they are still produced by the Rostov-on-Don plant.

Learning from failure

The Ka-25F and Mi-24 competition taught the Kamov designers that they should not stick too closely to the tactical and technical specifications initially submitted by the ordering party, but look to the future and develop a helicopter in accordance with requirements that may arise. The loss of the competition was a difficult experience, but it pushed the Kamov team to analyse what an ideal combat helicopter should be. In the following years, the Kamov designers developed, on their own initiative, a series of combat helicopter designs.[1]

In 1969, the V-50 project was created, which was to be a helicopter reaching a speed of almost 216kts (400km/h), made in a tandem twin-rotor configuration and with a very narrow fuselage. Two TV3-117s were to power the helicopter. The crew of the V-50 consisted of two pilots, with eight soldiers sat behind them, and weapons hung on the outer pylons. The V-50 was to be a universal helicopter, intended for both the army and the navy. Depending on the specific task, it was to take removable modules for reconnaissance, search, anti-submarine, anti-tank missile guidance, and more. A movable cannon would be installed under the helicopter's nose in the army variant, whereas the navy version would have a radar fitted. The V-50 project was developed intensively; even a series of wind-tunnel experiments were carried out. The design of the V-50 was led by Igor Erlikh, who had designed helicopters at the Alexander Yakovlev design bureau, including the tandem twin-rotor Yak-24.

The V-50 brought an understanding of the importance of the dynamic characteristics of the helicopter to the future Ka-52. The first acquaintance with the TV3-117 engine was also useful. The significance of the V-50 also lies in the fact it drew the navy's attention to the need for a seaborne-landing helicopter, which resulted in the creation of the Ka-29 *Helix-B* a few years later.

---

1 The whole story of the projects preceding the Ka-50 and Ka-52 was told to me by the Kamov deputy general designer Venyamin Kasyanikov in March 1993.

Above: Another idea for the V-100 was a high-speed single-seat fighter helicopter with combined propulsion. (Kamov)

Right: First on the right is 36-year-old Sergey Mikheyev, the newly appointed (in 1974) head of the Kamov design bureau. Next to him is Anufry Bolbot, Deputy Minister of the Aviation Industry of the USSR, and the three leading designers of the Kamov company: Igor Erlikh, Mark Kupfer and Nikolai Pryorov. (Kamov)

The next step was the V-80 project in 1971 – the first Kamov design of a purely combat army helicopter, without transport capacity. It still had the less powerful GTD-3F engines, but the fuselage was slim and light, unoccupied by the landing cabin; take-off weight did not exceed 7.5 tonnes. For the first time on Kamov helicopters, the landing gear was retracted. The crew consisted of a pilot and a gunner sitting in tandem (the gunner at the rear) in a large unarmoured glass cockpit. The main purpose of the V-80 was to cover the ground troops on the battlefield, thus fighting the enemy's aircraft. For this purpose, the V-80 was armed with a fixed cannon and air-to-air missiles. The Kamov designers proved the advantages of the specialised helicopter, especially its enhanced performance. However, they did not convince the military, which was constantly influenced by the idea of multitasking.

In discussions with the military, the designers were constantly told their proposals did not offer a significant advantage over the existing Mi-24. This pushed them to prepare the V-100 compound helicopter project in the mid-1970s. It had two side-by-side rotors at the wing tips and a small pushing propeller at the rear fuselage that allowed the helicopter to reach speeds of 216kts (400km/h).

This was the first design of the future Ka-50. It was decided from the very beginning to have one pilot and two co-axial rotors. (Kamov)

The V-100 had extremely powerful armament: 4,000kg of missiles, rockets and bombs on eight underwing and fuselage pylons and two movable cannons. The V-100 was intended not only for the army but for the navy as well. The navy variant had foldable wings, to facilitate parking on a ship, and was armed with two Kh-25 anti-ship missiles. The V-100 proposal was also not accepted by the military.

However, the V-100 project did influence today's Ka-52 design, with powerful weapons and crew ejection seats but, above all, an efficient integrated flight and targeting system. While working on the V-100, the designers thoroughly analysed the weapons and avionics that existed in the USSR at the time and familiarised themselves with prospective studies by various companies. Among other things, they gained knowledge of the Vikhr ATGM developed by KBP (Konstruktorskoye Byuro Priborostroyeniya, Instrument-building design bureau) of Arkady Shipunov in Tula and the Shkval optical-TV sighting sensor by the Zenith plant in Krasnogorsk, which turned out to be crucial in the history of the Ka-50 and Ka-52.

The assignment

In December 1976, the competition for the United States Army's advanced attack helicopter was settled. Hughes Helicopters' (then McDonnell Douglas, and today Boeing) AH-64 Apache was selected, the prototype of which was first flown on 30 September 1975. By 2020, 2,400 AH-64s had been produced, entering service in many countries.

The Soviets followed, and in the same month, on 16 December 1976, they ordered a similar helicopter from the Mikhail Mil design bureau, formally known as Mil Moscow Helicopter Plant (Moskovsky Vertolyotnyi Zavod, MVZ). After the death of Mikhail Mil on 31 January 1970, it was led by Marat Tishchenko. The Kamov company, formally known at that time as Kamov Ukhtomskaya Helicopter Plant (Ukhtomsky Vertolyotnyi Zavod, UVZ; Ukhtomskaya is a train station in Lyubertsy town outside Moscow), headed by Sergey Mikheyev after the death of Nikolai Kamov on 24 November 1973, initially did not participate in this project. Kamov was quite heavily loaded with the trials of the new Ka-27 ship-based helicopter. However, as soon as the Kamov designers learned about the new task assigned to Mil, they offered to develop another helicopter of this class. Since more options never hurt, the air force, together with the Soviet Ministry of Aviation Industry, announced a competition between them.

Nikolai Bezdetnov, who, for the first time on 17 June 1982, lifted the first Ka-50, '010', into a hover. (Kamov)

However, for the military, the competition was treated as a formality – not so for the Kamov designers. They knew that if two helicopters were similar, the Mil offer would be chosen out of habit – for decades the Soviet Air Force had been using armed Mi-4 and Mi-8 transport helicopters; the Mil team developed the first Soviet Mi-24 combat helicopter. For a Kamov helicopter to be selected, it would have to be a whole class better. They also knew that on an engineering level, they would not be able to beat their competitors from Mil – both of them used the same technologies and had studied at the same schools. So, they could only win with a completely new concept.

At this point, the Kamov team's several dozen years of experience began to pay off, particularly the ten years of dealing with army helicopters on paper. At the time of preparing the preliminary design, the helicopter was already fully thought out. As they had not built an attack helicopter before, they were not attached to any specific design solutions, which led to a different approach by both teams in the competition. The Mil design bureau was preparing its 'izdeliye (product) 280' helicopter (the later Mi-28), under the slogan of a deep modernisation of the Mi-24: 'We are removing the transport cabin, upgrading the weapons and equipment – and we get a new specialized combat helicopter at a low cost.' The Kamov design bureau took a different path, preparing, from scratch, a helicopter that would perform its tasks in the best way possible.

By the end of 1977, the preliminary design of the Kamov combat helicopter was ready; the subcontractors for the helicopter's equipment and armament were selected. The helicopter resembled the above-described V-80 design of 1971 and kept the same designation of V-80, also known as 'izdeliye 800'; many years later, in 1987, the helicopter received the designation Ka-50. The design was presented to the air force for approval in December 1977. The ordering parties were convinced, and the following months were spent agreeing on more detailed characteristics of the helicopter and its weapons with the air force, as well as building a full-size mock-up of the V-80. In May 1980, a joint commission of the air force and the Ministry of Aviation Industry approved the mock-up and, in August 1980, Kamov was awarded an order for two V-80 test helicopters; Mil received an order for two Mi-28s.

The test helicopters were made by the Kamov experimental production facility located at the Ukhtomskaya station in Lyubertsy near Moscow. In those days in the USSR, it was the workshops of the design bureau that built test examples of new aeroplanes and helicopters. It was not until much later, after test flights, that production was launched in the serial production plant. As a note, the production plant was chosen quite arbitrarily by the Ministry of Aviation Industry; Kamov helicopters have been series-produced throughout their history in Ulan-Ude, Kumertau, Orenburg and Arsenyev.

Key choices in the Ka-50 helicopter concept

A strategic decision in the development of the V-80 combat helicopter, the future Ka-50, was to give up the two-person crew typical of combat helicopters around the world; the helicopter was to be managed by only one pilot. The most important benefit of such a solution was the lighter weight compared to

Yevgeny Laryushin, who made the first full-circuit flight on Ka-50 '010'. On 3 April 1985, he died during a test flight in this helicopter. (Kamov)

a two-seat helicopter with the same characteristics. Of the 3,307lb (1,500kg) saved, about half was spent on increasing the firepower (a heavier and more effective cannon), increasing the combat survivability (739lb [355kg] of armour) and simplifying the maintenance.

In addition, with a one-person crew it was easier to facilitate the egress of the pilot using an ejection seat – a solution the designers considered necessary for a combat helicopter. The Zvezda team, led at the time by Guy Severin, made a K-37-800 ejection (or, more precisely, 'pulled-out') seat for the Ka-50.

The one-person crew design was proposed after analysing the work of the pilot and weapons system operator (WSO) in the Mi-24 helicopter. During an approach to a target, performed at a very low altitude, the WSO remained unnecessary, especially as only the pilot had a cartographic indicator in the cockpit. However, during an attack, all the tasks were performed by the WSO operating the targeting system. When attacking from a distance of 4km (this is the range of the Mi-24's anti-tank missiles), to visually observe the target, the helicopter must be 35–70m above flat terrain, or at least 100m above undulating terrain. In this position, piloting the helicopter does not pose any particular problems that would require the presence of a second crew member. Based on this, the Kamov designers decided to combine the functions of a pilot and a WSO into a one-person role.

Of course, the image above is somewhat simplified. The pilot workload in a single-seat helicopter would be large, if it was not for the high levels of automation in the helicopter equipment. In the Mi-24, where the anti-tank missiles were command-guided and flew at a low speed (170m/s for Falanga ATGM on the Mi-24D and 350m/s for the Shturm on the Mi-24V), an operator was needed to keep the sight crosshair on the target for a long time. Ten years later, the Kamov designers decided that, with new equipment and weapons, a single pilot could handle the operation. At this point in the helicopter's history, the experience with the V-100 project and the designers' resulting knowledge of the latest equipment and weapons systems developed in the USSR at that time came in handy. The Shkval (squall) targeting system coupled with the Vikhr (whirlwind) ATGM, then developed specially for the Su-25T single-seat attack aircraft, was adopted for the Ka-50.

The avionics of the Ka-50 helicopter were integrated into one PrPNK-80 Rubikon flight, navigation and targeting system prepared by the Elektroavtomatika company from Leningrad (today St Petersburg). The basic aiming system was the I-251V Shkval-V electro-optical sight by Zenit from Krasnogorsk (which, among others, produced the Zenit photo cameras popular in the USSR), intended for searching and automatic tracking of small moving targets, like tanks and helicopters. The Shkval had three sensors coupled together: a TV camera for target observation and tracking, a laser rangefinder and target marker and a laser-beam guidance system for the Vikhr ATGM. After detecting the target, the pilot would mark it on the TV screen and turn on the tracking system, which synced with the image of the target and continued to follow it. The same tracking system then aimed the laser beam at the target; the Vikhr missile would fly towards the target in the beam. Everything happened without the participation of the pilot. The Vikhr could fly at a speed of 610m/s, much faster than the previous ATGMs, so the helicopter did not have to stay within the anti-aircraft weapon's range for a long time.

The first Ka-50 prototype, '010', in its initial shape with a short tailfin and a high-wing wedge angle. (Kamov)

The idea to give up the second crew member was also supported by Kamov's own experience with ship-based helicopters. Years ago, when working on the Ka-25 naval helicopter, the navy required that it had only one pilot. As the pilot load on a ship-borne anti-submarine helicopter is greater than that on a land-based helicopter, the Kamov helicopters always featured a very high degree of systems automation.

Another strategic decision was the choice of the helicopter's rotor configuration. When working on the V-50 and V-100 projects, the designers analysed twin-rotor configurations: tandem and transverse. However, in defining the shape of the V-80, they stopped at their traditional two co-axial contra-rotating rotor arrangement. The most significant reasons for Kamov's choice of the co-axial rotor system were its known advantages over a single main-rotor and tail-rotor system.

Greater utilisation of engine power in the hover and low-speed flight gave the helicopter a better climb rate and enabled hot and high operations. The aerodynamic symmetry of the rotors and the small moment of inertia of the airframe in relation to the vertical and transverse axes meant better manoeuvrability and no vibration, which is an invaluable aid when aiming from a cannon and unguided rockets. The Ka-50 can take a position to attack in less time and in a smaller space than its competitors. For an army helicopter, a small silhouette is also important, as it makes the helicopter more difficult to detect and hit. This was later confirmed during the 1980s Afghanistan War, where 30 per cent of the losses of the Mi-8 and Mi-24 helicopters were caused by damage to the tail boom and its transmission, or to the tail rotor itself. There is no tail rotor on the Ka-50, and the tail boom is not a loaded element of the structure; in one of the flights, a Ka-50 lost its tail and returned to the base without any problems.

Another stimulus for such a choice was that, in 1977, the Ka-27 helicopter tests were ending, during which the TV3-117 engines, VR-252 gearbox and co-axial rotors were refined. The first idea was, as before in the Ka-25F project, to use this module from Ka-27 in the new Ka-50. However, during development, the requirements for the army helicopter rotor system became increasingly different from those implemented in the Ka-27. Finally, a new (or rather, deeply modernised) VR-80 gearbox and new rotors were designed for the Ka-50.

Under tests

On 17 June 1982, Nikolai Bezdetnov carried out the helicopter's ('800-01' bearing side number '010') first hover off the ground; on 27 July, Yevgeny Laryushin made the first circuit flight in it. The '010'

Ka-50 '010' has a fake co-pilot window and a painted-on cargo bay door. In this configuration, the helicopter has a higher tailfin and altered wing. (Kamov)

prototype was intended to evaluate the basic flight handlings and systems of the helicopter. It did not have any weapons yet, including a cannon; it also did not have an ejection seat for the pilot. The engines were temporary – TV3-117Vs. In the course of the tests, the shape of the wing and the tail were changed.

The '010' helicopter crashed on 3 April 1985, killing pilot Yevgeny Laryushin. The helicopter was performing a dynamic descent, failed to get out of it and hit the ground. Critics of the Ka-50 accused the helicopter of having a 'congenital defect' in the co-axial system, causing a clash of the upper and lower rotor blades during dynamic manoeuvring. However, the actual cause of the crash was the pilot exceeding the permissible limits specified for the helicopter. Nevertheless, the designers changed the lifting system of the helicopter after the crash, increasing the distance between the rotors by 120mm.

The second Ka-50, '011', already had the Shkval aiming system and Merkury low-light level TV (LLLTV) sight. A cannon is attached to the fuselage on the right side. (Kamov)

The third Ka-50 prototype, '012', has mock-ups of the daytime Shkval and night-time Mercury sights in its nose. A large box suspended under the outer underwing pylon houses the measuring equipment; next to it are cameras recording the movement of the rotor blades. (Piotr Butowski)

Previously, on 16 August 1983, Yevgeny Laryushin had taken off for the first time in the second helicopter, '800-02' (side number '011'), which was intended for testing equipment and weapons. The helicopter already had a starboard gun, a larger set of equipment and more powerful TV3-117VMA engines. During the trials, the Shkval aiming system and the Mercury low-light level TV (LLLTV) sight were installed on the '011'. During the tests, the aircraft accumulated 620 flight hours and fired about 100 anti-tank missiles. The V-80 preliminary trials were completed in July 1984.

The trials of the V-80 were not trials in themselves. Their main focus was still the comparison with the competing Mil Mi-28 helicopter. The dispute over the priority between the Mi-28 and Ka-50 continued for many years. It died down only when, in 2005–10, both helicopters received large orders, and everyone was mollified.

From December 1979, the Soviets fought in Afghanistan, and the issue of the rapid deployment of a new combat helicopter became urgent. The first large meeting on this subject was convened by the Soviet Air Force Main Command and the Ministry of Aviation Industry in October 1983. Based on the opinions of the Ministry of Defence (MoD) institutes, aviation systems institute GosNIIAS and aerodynamics institute TsAGI, as well as those of the pilots, the V-80 was recognised as superior. This decision was formalised in December 1984, when all institutes signed a resolution on the selection of the V-80 as a new combat helicopter for the Soviet Army, with more V-80s ordered for state evaluations.

The third test helicopter '800-03', side number '012', took off in December 1985. Similar to '011', a mock-up of the Mercury LLLTV sighting device was installed on it. Then there was a long break, with the development of '800-04', marked '014', not starting until April 1989. Despite the official

The first public presentation of the Ka-50 during the Mosaeroshow in Zhukovsky in August 1992. There, '012' flew in front of the audience in the Kamov helicopters column but was not available at the static exhibition. (Piotr Butowski)

Marxist–Leninist doctrine, the Soviets were superstitious, and the number 013 was omitted. Finally, in June 1990, the '800-05' (or '015') took off for the first time; it was the last of the five helicopters produced by Kamov's workshops in Lyubertsy and the standard for series production. It had a complete set of equipment, including self-defence devices, and – for the first time – an ejection seat. The '800-05' is one of the most popular Ka-50s. It appeared in the feature film *Black Shark* and has participated in numerous international airshows. In 2000–01, it took part in the operation in Chechnya (repainted, with the side number '25'). The sixth and seventh test V-80 helicopters were intended for ground tests.

However, back in 1984, the Mil lobby was much stronger within the Ministry of Aviation Industry, despite the military being in favour of the Kamov helicopter. In February 1984, right after the Mi-28 successfully completed the first stage of state tests, the Minister of Aviation Industry signed an order to start its series production in Arsenyev, with a deadline to build the first serial helicopter in the second half of 1987. Obviously, this command was not executed. Regardless of previous decisions, the ministry also pushed through the comparative tests of both helicopters under the same conditions.

Comparative trials of the V-80 (Ka-50) and '280' (Mi-28) helicopters began on 18 September 1985 at the Smolino training ground. They showed that the V-80 exceeded the combat effectiveness of the '280' helicopter due to better performance (especially in 'hot and high' conditions – in the mountains and at high temperatures), improved survivability over the battlefield and more effective armament. According to military pilots, the V-80 had lower vibrations, was more stable in strong cross or rear winds, had better manoeuvrability (it could make a tight flat turn at 200–220km/h) and a higher climb rate.

However, the key question was whether the electronics would be able to replace the second crew member and whether the V-80's single pilot with the Rubikon flight, navigation and targeting system with the Shkval sight would be able to perform the same tasks as the pilot and WSO of the 'classic' Mi-28 helicopter with the 9K113 Shturm-V anti-tank system. The tests were accompanied by constant breakdowns of the equipment, and the Ka-50 more often refused to obey. Nevertheless, the Ka-50 was able to demonstrate hitting a target with a Vikhr missile at a distance of 5 miles (8km), while the Ataka missile launched from the Mi-28 only reached 3.3 miles (5.3km).

In October 1986, the leading institutes of the Soviet Union's MoD issued their opinions: the victory 'on points' was won by the V-80. However, the evaluation committee, being between the hammer and the hard place, recommended to continue both programmes, albeit with different priorities. Following this, on 14 December 1987, the Central Committee of the CPSU and the Council of Ministers of the USSR adopted a major resolution on helicopters. It was ordered to prepare the final version of the V-80 helicopter, designated V-80Sh1, and to start the serial production with military designation Ka-50 at the plant in Arsenyev near Vladivostok, which was then producing Mi-24 helicopters for the USSR military aviation. Together with this, it ordered to continue the development of the Mi-28 for export, with production planned in Rostov-on-Don, which was then producing Mi-24s for export. This resolution had future importance, as it gave rise to another version of the helicopter, the two-seat combat-reconnaissance V-80Sh2 (later Ka-52).

In 1988–90, all four V-80 test helicopters were actively flying to refine the rotors system, flight control system and landing gear. The '012' and '015' helicopters were used to assess flight performance, while '011' and '014' were harnessed for weapons testing, including the gas-dynamic stability of the engine while firing a cannon and missiles, as well as ensuring electromagnetic compatibility of the equipment. Some of the tests, such as specialised equipment and weapons, were carried out on an experimental LL-800 (Letayushchaya laboratoriya, flying testbed) helicopter converted from a Ka-27.

In September 1990, the Ka-50 entered state qualification tests; '014' and '015' participated in the first stage. In January 1992, they were joined by '018', the first pre-series helicopter. From February 1992 to December 1993, the so-called 'second stage' of state tests, comprising the assessment of the combat effectiveness of the helicopter, took place. The document formally confirming the completion of the helicopter qualification tests was approved by the commander-in-chief of the air force, Pyotr Deynekin, on 2 December 1993, and by the commander-in-chief of the ground troops, Vladimir Semyonov, on 21 January 1994. Finally, on 28 August 1995, President Boris Yeltsin signed a decree officially commissioning the Ka-50.

'015', which made its first flight in June 1990, received a complete set of equipment, including the first Ka-50 ejection seat, and was a guide for the series production. (Kamov)

This is how the Ka-50 was imagined in the West before it was publicly shown. This is a drawing from the *Soviet Military Power 1987* brochure by the US Department of Defense (DoD). The theory is that the DoD knew very well that it was a single-seat helicopter, but did not reveal this knowledge. All the drawings published at that time show the helicopter from the front, so the cockpit was not visible. (US DoD)

First information in the West

The first information about the new Soviet combat helicopter named *Hokum*, along with a small but generally correct sketch, was published in the annual report of the US Department of Defense, *Soviet Military Power 1985*; the Mi-28 *Havoc* was disclosed a year earlier. The drawing of the helicopter was accompanied by two figures: a speed of 350km/h and an operating radius of 250km. In the following years, the illustrations became more and more precise. In 1989, the first poor-quality photo of the helicopter appeared in the West. Various designations were assigned to it in publications: Ka-34, Ka-37, Ka-136 and sometimes Ka-41. For example, the *Jane's All the World's Aircraft 1991-92* yearbook wrote: 'OKB designation Ka-136 and service designation Ka-34 unconfirmed.' It was only in January 1992 that Kamov released the first clear photographs of the helicopter, and in the summer of 1992 – ten years after the helicopter's first flight! – the Russians announced its real designation: Ka-50.

Why was the Ka-50 hidden for so many years, while other Soviet aircraft – under the policy of 'glasnost' – were shown to the world? Because it was an unusual helicopter, not only in its design, but in its concept. The most concealed element was that it was a single-seat helicopter.

On the first two prototypes, '010' and '011', a second pilot's cockpit was painted, as well as a small window and a door of the (non-existent) transport cabin. It was similar to the Su-25T attack aircraft with the same weapon system: its first prototypes had a second cockpit painted in to suggest this was nothing new – merely a Su-25UB training aircraft.

Chapter 2
Ka-50 in production and service

In 1990, the Military and Industrial Committee of the Council of Ministers of the USSR commissioned the construction of a preliminary series of 12 Ka-50 helicopters. Along with this, Kamov began to transfer the production documentation to the Progress plant in Arsenyev. The first Ka-50 produced there, number '018', took off on 22 May 1991, piloted by Anatoly Dovgan. The production variant of Ka-50 was given the internal designation 'izdeliye 800.05'.

Further production slowed down due to a lack of financing. In 1992, the plant constructed two more helicopters, '020' (later repainted to '024') and '021' (later converted into Ka-52 '061'), which were sent for testing at Kamov. In 1993, three helicopters were made – '20', '22' and '24' – and, in November 1993, the first two of them were handed over to the armed forces centre in Torzhok. In the following years, the plant produced one or two helicopters every few years. The last Ka-50, '28', left the Arsenyev factory in November 2009. However, it did not last long. Operation of Ka-50s in the Russian Air Force ended in November 2012.

A total of 18 or 19 Ka-50 helicopters were manufactured; it is hard to be sure, as they were often repainted and not all were finished. Some unfinished Ka-50s were later used in the construction of the two-seat Ka-52s. At its peak, Russian Army Aviation was using only six Ka-50s, all at the 344th Combat Training and Aircrew Conversion Centre in Torzhok. It was planned to deploy Ka-50s to an operational aviation unit, the 319th Helicopter Regiment based at Chernigovka, 37 miles (60km) from Arsenyev (the same unit was the first equipped with Mi-24 combat helicopters in 1971), but this never happened.

On 17 June 1998, at 2158hrs, Ka-50 number '22' crashed at the centre at Torzhok. Major General Boris Vorobyov, the commander of the centre, who was awarded the title of 'Hero of the Russian Federation' in 1996 for the Ka-50 trials, was killed in the crash. During the flight, at an altitude of 165–230ft (50–70m), the rotor blades broke off and the helicopter fell onto the runway. According to the investigation commission, the crash was caused by a significant exceedance of the permissible flight envelope, which led to a clash of the upper and lower rotor blades during manoeuvring.

Anatoly Dovgan is the pilot who, on 22 May 1991, flew '018', the first Ka-50 produced by the Progress plant in Arsenyev, for the first time. Here, Dovgan is sitting in the cockpit of a Mi-24. (Archives)

Kamov Ka-50 Hokum-A Helicopter Specifications

Engines: 2 x Klimov VK-2500 rated at 1,790kW (2,400shp) each

Principal dimensions:
Rotor diameter: 47ft 4in (14.43m) upper, 47ft 5in (14.46m) lower
Maximum length, rotors turning: 52ft 4½in (15.96m)
Fuselage length: 49ft 4in (15.03m)
Wingspan: 24ft 1in (7.34m)
Maximum height: 16ft 3ins (4.95m)

Undercarriage:
Type: Retractable with twin nose wheels and single main wheels.
Wheel base: 16ft 1in (4.91m)
Wheel track: 8ft 9in (2.67m)

Weights:
Empty: 16,958lb (7,692kg)
Normal take-off: 21,605lb (9,800kg)
Maximum take-off, for combat: 23,810lb (10,800kg)
Maximum take-off, for ferry flight: 26.235lb (11,900kg)

Performance:
Maximum level flight speed: 167kts (310km/h)
Cruise speed: 146kts (270km/h)
Service ceiling: 18,045ft (5,500m)
OGE hover ceiling, normal weight: 13,123ft (4,000m)
OGE hover ceiling, maximum combat weight: 11,483ft (3,500m)
Maximum climb rate at sea level, 70kts (130km/h) speed, normal weight: 16m/s (3,150ft/min)
Practical range at normal weight, 5% reserve: 283 miles (455km)
Ferry range, 5% reserve: 721 miles (1,160km)

Ka-50 '018' was the first helicopter produced by the Arsenyev plant. (Piotr Butowski)

Above: For many years the Ka-50, and then the Ka-52, fought for recognition with its competitor, the Mil Mi-28 helicopter. Here is Ka-50 '26' from the 344th Combat Training and Aircrew Conversion Centre in Torzhok, followed by two Mi-28Ns. (Piotr Butowski)

Right: This is '020' with the c/n 3538053201003, the third example of the first production batch. H318 is the number the helicopter was given at Le Bourget in 1993. (Piotr Butowski)

Despite the number '024', this is the same '020' as in the previous picture; the number was repainted in 1995. (Piotr Butowski)

This is '20', the first helicopter of the second production series, built in October 1993 and delivered to the centre in Torzhok. In the photo, the helicopter is undergoing an overhaul at the Kamov workshop in Lyubertsy in the late 1990s. (Kamov)

Left: '24' was the fourth helicopter of the first production series, which later, in 2000–01, participated in the operation in Chechnya, and since 2011 has served as a technical aid at the Air Force Academy in Voronezh. (Piotr Butowski)

Below: Ka-50 '25' is the first helicopter partially equipped with the L370 Vitebsk self-defence suite. L370-2 missile approach-warning sensors are placed on the sides of the front fuselage, while decoy launchers are installed in the wingtip fairings. (Piotr Butowski)

This is the same Ka-50 '25' with elements of the Vitebsk self-defence suite. Under the fuselage are places prepared for L370-5 infrared jammers, but the sensors themselves are not there yet. (Piotr Butowski)

Right: Ka-50 '26' is c/n 3538053003002, the second helicopter of the third (and last) production batch made in 2000, serving in the crew conversion centre in Torzhok. (Piotr Butowski)

Below: In November 2009 (nine years after the previous Ka-50), the last two Ka-50s were produced: '27' (c/n 3538054903004) and '28' (c/n 3538054903005). They flew for a very short time, because the Ka-50s ceased being operated in November 2012; '27' accumulated 221 flight hours. (Piotr Butowski)

Major General Boris Vorobyov (first from right), commander of the 344th Combat Training and Aircrew Conversion Centre and enthusiast of the Ka-50 helicopter, was among the centre's pilots. On 17 June 1998, Vorobyov was killed in the Ka-50 crash. (Piotr Butowski)

Left: There are few photos of black Ka-50 '22' with a shark jaw painting, as it was painted in late 1996 or early 1997, and on 17 June 1998, Boris Vorobyov crashed this helicopter. (Piotr Butowski)

Below: The main structural element of the Ka-50 (pictured) and Ka-52 fuselage is a box beam with fuel tanks and cannon ammunition inside. Attached outside are equipment modules covered by light external panels. Such a configuration, although heavier than the classic one, enables better use of the internal volume and facilitates access to the equipment during maintenance. (Kamov)

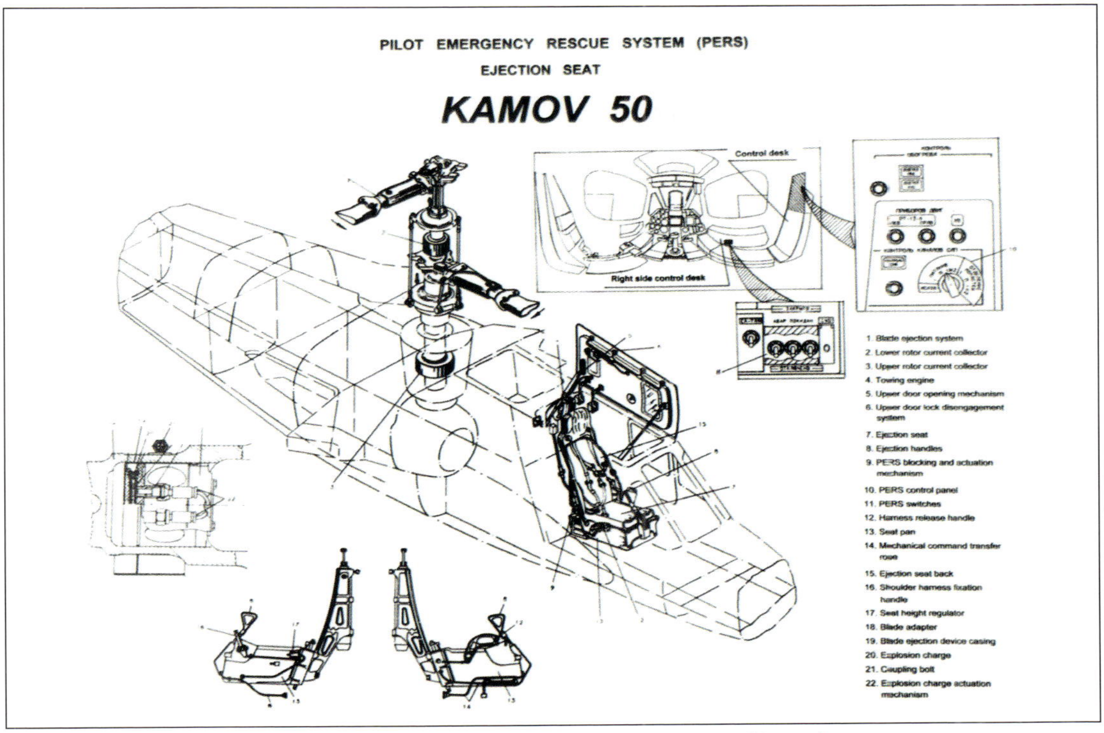

Above: Ka-50's pilot's rescue system with the Zvezda K-37-800 ejection seat. (Kamov)

Below left: The K-37-800 seat as presented at Farnborough International Airshow 1996. (Piotr Butowski)

Below middle: A mock-up of the Ka-50 cockpit, made long before the first flight of the helicopter, in the Kamov museum in Lyubertsy, 1993. Currently, neither the museum nor the mock-up exists. (Piotr Butowski)

Below right: Cockpits of various Ka-50s differ from each other. This is the most typical cockpit, seen here in '021'. In the centre of the instrument panel, there is a TV screen of the Shkval-V sight, with a head-up display (HUD) above. The small windows on the sides of the HUD are interesting. These are the sensors of the Obzor-800M helmet-mounted system for initial target indication for the Shkval sight. (Piotr Butowski)

Control stick and side instrument panels in Ka-50 '021' pilot's cockpit. The red handles activate the ejection seat. The cardboard box in the place of the pilot's left foot is not standard equipment! (Piotr Butowski)

Left: The ceiling of the Ka-50's cockpit is glazed and equipped with a mirror for rearward observation. (Piotr Butowski)

Below left: The cockpit of Ka-50 '020' has been adapted for the use of night-vision goggles (NVGs) by the pilot. (Piotr Butowski)

Below right: The large rectangular window of the Shkval-V TV sight under the Ka-50 nose; the window has a doormat. The small square window on the left is the L140 Otklik laser-warning sensor. (Piotr Butowski)

Above left: The Shipunov 2A42 30mm cannon is mounted on the starboard side of the Ka-50's fuselage. The hydraulic drive moves the barrel -37/+3.5° in elevation; aiming in azimuth is affected by positioning the helicopter itself. (Piotr Butowski)

Above right: A B-8V20 launcher with 20 80mm rockets and a UPK-23-250 gun pod with twin-barrel 23mm cannon and 250 rounds. (Piotr Butowski)

The six-tube UPP-800 launcher for Vikhr ATGMs has an adjustable position. Its nose can be tilted downwards by 10° to keep the missile within the laser beam aimed at the target. (Piotr Butowski)

A Ka-50 fires 80mm unguided rockets. (Kamov)

A Ka-50 launches a Vikhr anti-tank guided missile. (Kamov)

Ka-50 in Chechnya

Although the Russian press reported as late as 1999 that two Ka-50s had been transferred to Mozdok near the Chechen border, in reality, they had not moved from their permanent base in Torzhok. Most likely, the main reason for the reticence of the Russians was the tender in Turkey, which was then entering its culminating phase. Directing the Ka-50 against rebels in Chechnya would not help its image in Islamic Turkey.

On 27 December 2000, a pair of Ka-50 helicopters, '24' and '25', were transferred from Torzhok to the Caucasus. The single-seat Ka-50 was not a 'free hunter' and was primarily used to make precise strikes at targets indicated from the outside. Its combat capabilities could only be fully utilised when it was operating in a group along with a combat control and target acquisition helicopter.

Therefore, to test the capabilities of the Ka-50 in real combat conditions, an experimental BUG (Boyevaya Udarnaya Gruppa, combat strike group) was formed, in which, together with two Ka-50s, there was a converted Ka-29VPNTsU helicopter (side number '35') adapted to cooperate with Ka-50s as a combat control helicopter; VPNTsU stands for Vozdushnyi Punkt Navedeniya, Tseleukazaniya i Upravleniya (airborne homing, target indication and control post). The Ka-29VPNTsU received the Rubikon system (the same as in the Ka-50) with Shkval sight and coded communication equipment. The helicopter also received a 2A42 cannon (it is optional for the Ka-29), as well as hot-air mixers at the engines' exhausts and flare launchers for self-defence. The BUG force was also assigned several Mi-24s for cover. For the expedition to Chechnya, the helicopter numbers, as well as the nationality markings, were painted over.

The group operated from Grozny North Airport. First, the Ka-50 pilots flew the Mi-8 and Mi-24, getting to know the area of operations. Later, a single Ka-50 was included in the Mi-24 combat group and, finally, the BUG entered the action. As experience was gained, the Mi-24 cover was gradually abandoned. On 6 January 2001, for the first time in history, the Ka-50 opened fire on a real enemy. The separatists camp in the mountain pass near the village of Tsentoroi was destroyed with two

Above left: In 2000–01, Ka-50 '25' (previously known as '015') took part in the operation in Chechnya. Here, it is presented in the Kamov's trials site in Shcholkovo (Chkalovskaya) near Moscow shortly after returning from Chechnya. (Piotr Butowski)

Above right: Two Ka-50 combat helicopters were supported in Chechnya by this Ka-29VPNTsU homing, target indication and control helicopter. (Piotr Butowski)

Right: A Shkval sight mounted under the Ka-29VPNTsU's front fuselage. (Piotr Butowski)

Vikhr missiles fired from a distance of 3km. The helicopters returned to Torzhok on 16 February 2001. During the six-week operation in Chechnya, helicopters destroyed 30 targets (such as separatist groups, vehicles, mountain bases).

The Ka-50, acting together with the Mi-24, showed an advantage in terms of operational altitude, manoeuvrability, fire accuracy and the weight of the second salvo. The military collected many recommendations for the further development of the helicopter. The first conclusion became known many years earlier: combat helicopters must operate around the clock, not only during the day, so the development of the night-time Ka-50Sh needed to be accelerated. The then results of work on night helicopters in Russia, not only the Ka-50Sh but also the converted Mi-24 and Mi-8, were unsatisfactory; there were just a dozen upgraded helicopters in service at the time, most of them in combat in Chechnya.

Another recommendation was to increase the helicopter's survivability over the battlefield, in particular the strengthening of its armour. Russian attack aeroplanes and helicopters have always had strong armour. One of the Kamov designers said that 'until recently the military blamed us for unnecessarily overloading the helicopter, while after the fighting in Chechnya they demanded even more protection'. It was also a stimulus for the development of the new L370 Vitebsk self-defence suite;

in Chechnya, Ka-50s only had the L140 Otklik laser-warning sensor and UV-26 decoy dispensers. The military also applied for an improvement of hot and high performance; later, for this purpose, the Ka-50 and Ka-52 were fitted with uprated VK-2500 engines instead of the TV3-117VMA.

Night-time Ka-50 versions

In 1982, after the first experiences of the Soviet war in Afghanistan, a government resolution was adopted in the USSR recommending that all close-air support aircraft and helicopters could be operated at night. This gave rise to active work on low-light level TV and thermovision sensors, as well as lightweight millimetre-wave radars.

The results of the work were unsatisfactory for a long time. Back in the 1980s, a mock-up of the Merkury LLLTS camera was installed on the second ('011') and third ('012') prototypes of the Ka-50. However, after the trials conducted on Su-25T attack aircraft, on which the Merkury was also planned to be installed, the system was disqualified due to extremely low performance and high unreliability. For some time, the new Stolb (column) thermal imaging system by the Geofizika company was tested (also only in the form of a mock-up on the helicopter); it was also abandoned. Similarly unsuccessful were the tests with the Khod (move) thermal imager and the Kinzhal-V (dagger) radar, conducted on the Su-25T; they were also disqualified due to poor performance.

The gyrostabilised Samshit (boxwood) electro-optical (EO) payload turned out to be the most, or actually the only, successful one from this series of devices. It was developed by the Ural Optical and Mechanical Plant (UOMZ) in Yekaterinburg (until 1991 called Sverdlovsk) and is still produced in various variants today.

On 4 March 1997, the pilot Oleg Krivoshein made the first flight of the Ka-50Sh helicopter; it was the modernised first production Ka-50 '018' (later repainted to '18'). The same month, the helicopter was shown at the International Defence Exhibition (IDEX) 1997 in Abu Dhabi. The Ka-50Sh received the GOES-346 Samshit-50T night-sighting system installed in the fuselage nose; GOES stands for Girostabilizirovannaya Optiko-Elektronnaya Stantsiya (gyrostabilised opto-electronic station).

Night-time Ka-50Sh '018' in the configuration shown during MAKS 1997, with a Samshit-50T turret mounted below the standard daytime Shkval-V sight. (Piotr Butowski)

Above left: In 1999, the same Ka-50Sh '018' received an improved Samshit turret that housed a TV camera, thermal imaging camera, laser rangefinder and laser-beam riding for Vikhr ATGM. (Piotr Butowski)

Above right: The Ka-50Sh mock-up made from retired '014' in 1999 had a large 460mm GOES-330 aiming sensor and a small GOES-520 turret for piloting. (Piotr Butowski)

The Samshit-50T was a rotating ball, inside which sensors on a gyroscopically stabilised platform were installed: thermal imaging camera, laser rangefinder and laser-beam riding for Vikhr ATGM. Initially, the Samshit ball was placed in the helicopter nose above the regular Shkval-V window, but a few months later it was swapped: the Shkval was moved upwards and placed in the elongated nose tip, while the Samshit was mounted underneath. The helicopter was shown in this configuration for the first time during MAKS (Mezhdunarodnyj aviatsionno-kosmicheskij salon, International Aviation and Space Show) 1997 in Zhukovsky.

Ten years later, the sensors in the front of the Ka-50Sh '18' fuselage were reconfigured: the large Samshit ball (in the new version) remained at the bottom, but the Shkval at the top was replaced by a simple TOES-521 turret, intended for piloting, without the capability of target tracking and aiming weapons. The helicopter was shown in this configuration at Zhukovsky in 2007.

Above left: Another configuration of Ka-50Sh '18' (ex '018') sensors as presented in 2005. (Piotr Butowski)

Above right: The final configuration of night-time Ka-50Sh '18' sensors as presented in 2007, with targeting GOES-346 below and piloting TOES-521 in the nose tip. (Piotr Butowski)

Above left: A Saturn pod with French Thomson-CSF Victor thermal imaging sensor inside, suspended under the wing of Ka-50 '020' at MAKS 1995. (Piotr Butowski)

Above right: At Farnborough 1998, the Russians presented a mock-up of a new cockpit for the Ka-50 made by the Russkaya Avionika (Russian avionics) company; it was never realised. (Piotr Butowski)

Left: A helmet-mounted sight and display by GEC Marconi was proposed for Ka-50 at Farnborough 1998. (Piotr Butowski)

The same '018' was also the first Ka-50 on which an Abris navigational panel (named Kabris in the version for Kamov helicopters), made by the Transas (Kronshtadt from March 2000) company, was installed. It was a vectored digital chart coupled with a satellite navigation receiver; before that, the Ka-50 pilot had a paper map. Also on the '018' helicopter, the 'open architecture' of the avionics, with a data bus, was first implemented in the late 1990s; some instruments in the cockpit were replaced with liquid crystal displays.

In summer 1999, another variant of the night-time Ka-50Sh was shown, in the form of a mock-up made of the retired '014' helicopter. The equipment, including the new glass cockpit, was made by the Ramenskoye PKB design bureau. Two EO turrets were installed in the shortened front of the fuselage, the larger 460mm GOES-330 aiming system and a small GOES-520 turret for piloting.

In those days, Russia's cooperation with the West was quite close, and many Western systems were fitted to the Ka-50. For example, French Thomson-CSF Victor and Swedish Agema THV1000 thermal imaging sensors were housed by the first Samshit payloads. At the MAKS 1995 airshow, Ka-50 '020' was shown with a Saturn pod suspended under the wing, with the Victor FLIR inside. The system had two observation channels, wide channel (5.7° x 8.6°) and narrow channel (1.9° x 2.9°), capable of detecting a tank at a distance of 6–7km and aim at 4.5–5km. At Farnborough Airshow 1998, the Russians presented a mock-up of a new cockpit for Ka-50 made by the Russkaya Avionika (Russian Avionics) company and with a helmet-mounted sight and display by GEC Marconi.

Chapter 3
For export: the first attempt

While planning the future of the Ka-50 and Mi-28 helicopters in 1987, it was decided the first one would be used by the USSR Air Force, and the second one would be exported. In reality, this intention did not last long. In the mid-1990s, Russia was in a deep economic crisis, and the MoD stopped purchasing new equipment. Therefore, the main goal of all Russian defence industry companies, including Kamov, had become to secure contracts abroad.

In early 1992, the Russians launched a carefully programmed promotional campaign for the Ka-50. In January 1992, an international conference on fighter helicopters was held in London, during which Sergey Mikheyev gave a lecture on the Ka-50. He presented the helicopter idea, described the construction and revealed some good quality photos – but he had not even given the designation of the helicopter yet. Several more months passed, during which more news about the helicopter, including its name, the Ka-50, was submitted to the press. In August, in Zhukovsky near Moscow, the Mosaeroshow '92 exhibition (renamed MAKS from 1993) was held; the programme also included the Ka-50. However, the visitors saw it only from a distance. Ka-50 '012' flew over the airfield in the Kamov helicopters column; it was not displayed at the static exhibition.

Above left: Sergey Mikheyev, the Kamov company designer general, presents the Ka-50 at Farnborough in 1994. (Piotr Butowski)

Above right: Ka-50 '021' with the full set of its weapons as presented in Le Bourget in 1993. 'H317' is the number assigned by the organisers of the Paris Airshow. (Piotr Butowski)

The second helicopter performing at Le Bourget in 1993, Ka-50 '020', or H318, participated in the flight display. (Piotr Butowski)

An advertisement campaign was being prepared for Farnborough in early September 1992. This, however, did not quite work out. When the helicopter arrived in England on a transport aircraft, it turned out there was no one to look after it. The entire 18-person staff, including the pilot, had not been granted visas. The British authorities maintained this was because of the late submission of documents by the Russians; the Russians blamed Western competitors for deliberately blocking the helicopter presentation. The truth was probably somewhere in the middle.

The helicopter presented in Farnborough was Ka-50 '020', the third copy made by the Arsenyev factory. The unusual helicopter was also unusually painted. It was all black, with a tricolour Russian flag on the engine nacelles. At the tailfin, there was a drawing of a wolf's head and the name 'Werewolf', by which the Ka-50 was baptised for export. The modest painting was completed with a red triangle and the name 'Ka-50' on the tailfin; the logo of the Kamov design bureau was placed on the vertical plates at the tail plane, and the words 'Army attack helicopter' (in English) on the left side of the fuselage.

After Farnborough 1992, the Ka-50 became a regular guest at the world's airshows. Two Ka-50s performed at Le Bourget near Paris in June 1993. The first one, participating in the flight demonstration, was the same '020' that was previously at Farnborough; it also had the Le Bourget number 'H318'. The second helicopter, presented at the static exhibition, had the side number '021' and the exhibition designation 'H317'. Unlike '020', which had a werewolf painted on its tail, '021' had an image of a shark and the words 'Black Shark'.

Ka-50 '024' shown at Farnborough in 1996 is the same '020' as in the previous photo. Its number was changed in 1995. (Piotr Butowski)

In 1995, Ka-50 '020' had changed its number to '024'. It participated in most of the international exhibitions in the following years. In 1995, it was presented at the Langkawi International Maritime and Aerospace Exhibition (LIMA) in Malaysia; in 1996, it came back to Farnborough, and, in 1997, it was at Le Bourget once again (it was given the exhibition designation 'H347'). In early December 1996, at the Aero India exhibition in Bangalore, Ka-50 '024' was shown accompanied by the '061' prototype of the two-seater Ka-52 (this was before the Ka-52 made its first flight). In 1997, two Ka-50s, '18' (formerly '018') and

Ka-50 '020' tailfin with the 'Werewolf' painting. (Piotr Butowski)

Ka-50 '021' tailfin with the 'Black Shark' painting. (Piotr Butowski)

'22', were presented at IDEX in Abu Dhabi, while helicopter '25' (formerly '015') performed at the International Defence Industry Fair (IDEF) in Ankara, Turkey. Of course, the Ka-50 also participated in all subsequent MAKS exhibitions in Zhukovsky in Russia.

Back in 1993, a helicopter presentation in Algeria was scheduled; however, negotiations were suspended after deterioration of the country's internal situation. Slovakia, which received Russian military equipment as a settlement for the USSR's debts to the country, also expressed great interest in the helicopter. At the end of October 1996, at the airfield in Kuchyňa in Slovakia, Ka-50 '018' performed demonstration shootings for Slovak military commanders. Ultimately, however, Slovakia chose the S-300 anti-aircraft missiles and MiG-29 fighters over the Ka-50 helicopters.

For a short time in 1992–93, Kamov's representative in the West was Group Vector of Virginia Beach, US, but cooperation was interrupted after numerous disagreements between the partners. The Russians were just learning how to navigate the international military equipment market. There were many reports in the press that the Ka-50 was to be produced in the US, and that it was also about to enter the competition for a helicopter for the British Army. They were all journalistic ducks.

As mentioned, a Ka-50 helicopter played the titular role in the Russian feature film *Black Shark*, which was released in cinemas in early 1994. It was an action movie about the fight against a drug gang, in which the Ka-50 presented itself similarly to the helicopter known in the West from the 1983 movie *Blue Thunder*.

Turkish saga

For many, many years, efforts to find a buyer for the Ka-50 and then the Ka-52 abroad were unsuccessful, but there were many developments along the way. The most interesting was the ten-year tender for a combat helicopter for Turkey.

On 30 May 1997, Turkey began the request for proposals (RFP) process in the planned purchase of third-generation attack helicopters ATAK. Turkey was going to buy and produce 145 combat helicopters for its armed forces in 2002–10, for a total sum of US$3.5 billion. Officially, the tender concerned only an order of 50 helicopters; tenders for the next two lots of 50 and 45 helicopters were

The first two-seat Ka-52, '061', with side-by-side seats during the Turkish tender presentation; note the 'Ka-50-2' on the tail. (Kamov)

to be announced separately later, but it was obvious the winner of the first tender would have the next two almost in their pocket. The deadline for submitting proposals to the Turkish undersecretariat for defence industries (SSM) was 31 December 1997, and on that day all bids, including Kamov's bid, were submitted.

Previously, a Russian company participating in a tender for military equipment for a NATO member had been completely unimaginable, but, in the 1990s, it became possible. Moreover, the Russian MoD had no money, and a foreign order would be a lifeline for Kamov and the Ka-50/Ka-52 programme. Therefore, the idea of taking part in this tender interested the Kamov management, and later also the Rosvooruzheniye arms trade company (today Rosoboronexport). At the beginning of 1997, Sergey Mikheyev, within a large Russian delegation, visited Turkey and held a number of meetings with industry leaders and the MoD.

Sergey Mikheyev knew the Russian proposal would be an outsider in this tender. So, to win, it was necessary to present something non-standard, which (again, similarly to the competition with the Mi-28) would stand out among other offers. The Russians decided to join forces with Israel. During the Le Bourget airshow in June 1997, Mikheyev met with the management of Israel Aircraft Industries (IAI) and proposed the idea of joint participation in the Turkish tender. Kamov already had some experience working with IAI, which was in charge of marketing the civilian Ka-64 Sky Horse helicopter, a joint project between Kamov and Agusta of Italy, being a version of Ka-60 powered by General Electric CT7-2D1 and with a conventional tail rotor. This project was actively promoted in 1995–96, but then stopped because of funding shortcuts from the Russian side.

Negotiations on the terms of the Russian–Israeli partnership in the ATAK programme dragged on for several months. In the end, it was established that in the Russian–Israeli tandem, Kamov would be responsible for the helicopter, including the engines and the rotor system, while the IAI Lahav division was to prepare the avionics, including the cockpit, fire-control system, sensors and also the navigation, communication and electronic warfare devices. Thus, Kamov and IAI entered the Ka-52 into the ATAK competition and a long epic contest began.

As part of the preparation of the offer, the Russians conducted many demonstrations of the helicopter. It started in April 1998 with the presentation at Kubinka Air Base near Moscow. Several pilots from Turkey and Israel made familiarisation flights on two-seater Ka-52 '061' (guest pilots flew in the co-pilot's seat; Kamov test pilot Alexander Smirnov piloted the helicopter). Israeli pilots flew the Ka-52 again in September 1998. In October 1998, at the Lukhovitsy airfield near Moscow, Lior Parag, chief helicopter test pilot at IAI, made the first solo flight. The same month, Ka-52 '061' successfully underwent performance and flight tests at a military training ground near Antalya in Turkey, also with the participation of Turkish pilots. The helicopter received very good feedback, especially for its manoeuvrability and hover ceiling.

The second stage of the tests, consisting of evaluation of the helicopter in night flights and with the use of various types of weapons, began in August 1999. Two helicopters, Ka-52 and Ka-50, flew to

The glass cockpit for the Ka-52 proposed by the Israeli Elbit for Turkey, at Farnborough 2002. (Piotr Butowski)

Turkey. The participation of the Ka-50 was because the only then-existing two-seater version, '061', was more of a demonstrator than a combat helicopter – it did not yet have a weapon system. Therefore, the presentation of cannon firing, unguided rockets and Vikhr anti-tank missiles was made by the single-seat Ka-50. The pilots during these demonstrations were Colonel Sergey Zolotov from the 344th Combat Training Centre in Torzhok and Yuri Timofeyev from the Kamov company. In mid-October 1999, a delegation from the Turkish Armed Forces visited the Progress plant in Arsenyev to see the helicopter production line.

Originally, the Ka-52 helicopter with the side-by-side seats was submitted to the Turkish competition. However, in the spring of 1999, after the candidates had been presented, the Turks put forward an additional requirement – that the pilots should sit in tandem. Therefore, in the summer of 1999, another configuration was formed in the *Hokum* family: the two-seat Ka-50-2, with seats one in front of the other. Mikheyev decided to do the tandem two-seater version in the same way as the previous side-by-side Ka-52: by cutting the front fuselage on frame number 7 and replacing the front part with a new one. The rear seat was raised up compared to the Ka-50, and the front seat was moved forward and lowered. The crew was placed in an unusual way: the pilot was in the front and the weapons operator in the rear. Unlike the other versions of the helicopter, the Ka-50-2 crew did not have ejection seats.

In the Russian–Israeli variant intended for Turkey, the Ka-50-2 helicopter was named *Erdoğan*, which in free translation means 'born to fight'; this was in no way connected with the current (at the time of writing) Turkish president, Recep Tayyip Erdoğan, who was in deep opposition at the time. The small Ka-50-2 model was first shown in the summer of 1999 in Le Bourget. The full-size mock-up of the Ka-50-2 *Erdoğan* was first displayed at the IDEF '99 exhibition in Ankara, Turkey, in late September.

Turkey also demanded that some weapons be replaced with those compatible with NATO standards, including the French 20mm Giat 621 cannon with 700 rounds instead of the original Russian 30mm 2A42 cannon with 460 rounds. What's more, the cannon was supposed to fire in any direction around the helicopter, so it had to be mounted on a flexible carriage. The gun carriage was attached to the right side of the fuselage, and the cannon tilted under the fuselage and rotated 360° for firing. In the pulled-up position, the gun was fixed at the right side of the fuselage; in this position, it could also fire. The anti-tank armament was to include Russian Vikhr, Israeli Rafael NT-D or American Hellfire ATGMs. The Russian 80mm S-8 unguided rockets were supplemented with NATO 70mm rockets. The IAI Lahav division's avionics system, including glass cockpit, fire-control system, NavFLR electro-optical piloting/navigation turret, HMOPS EO targeting payload, helmet-mounted display and sight, as well as navigation, communication and electronic warfare systems, were planned for the Ka-50-2.

Developing the Russian–Israeli Ka-50-2 for Turkey, Kamov thought about using the ready tandem two-seat design for other customers; the designation of the Ka-54 was planned for it. However, these ideas did not go beyond the paper.

A full-size mock-up of the Ka-50-2 Erdoğan helicopter with tandem crew seats participating in the ATAK tender in Turkey at MAKS 2001. (Piotr Butowski)

Five helicopters were submitted to the Turkish ATAK tender. The first stage of the competition was resolved on 6 March 2000, when Turkish Prime Minister Bülent Eçevit announced the shortlist, which included the Helicopter Textron AH-1Z King Cobra, Agusta A129 International and Ka-50-2. The American Boeing AH-64D Apache Longbow was eliminated due to the high price and inability to meet the delivery start date in 2003, which was then a prerequisite for the contracting authority. The Franco-German Eurocopter Tiger was also unable to meet the deadline. In July 2000, the Turks announced the winner of the competition was the AH-1Z and entered into detailed negotiations. However, these did not bring results. The US did not agree to the transfer of licences for the production of on-board computers and software in Turkey.

The Russians tried to enter the void again, and during a visit to Turkey from Prime Minister Mikhail Kasyanov at the end of October 2000, they presented further bargaining chips. In addition to the helicopter, Turkey was offered joint production of tanks and armoured personnel carriers, investments in the Turkish defence industry and additional supplies of natural gas in 2001. As an offset, Kamov offered joint design and production in Turkey of the Ka-115 light utility helicopter. Finally, a reduction in the Ka-50-2's price was offered. In July 2002, the Turks resumed talks with the Russians. In 2003, the Russians were enthusiastic and announced the contract with Turkey was 'almost signed'. Admittedly, not only with them: talks were also underway with Bell Helicopter Textron at the same time. However, on 14 May 2004, the tender was unexpectedly cancelled. Prime Minister Erdoğan later said that 'during the tender procedures, cases of corruption in the government and the General Staff were revealed; therefore the tender was closed, but will be resumed in the near future.'

In August 2004, Turkey announced another tender for 50 combat helicopters, with an option for 41 more. The Russians, after a brief reflection, decided to participate again but on their own, without any Israeli companies. They felt they had enough experience and technology to not need help. They offered a lower price, approval to export the helicopter to third countries and added the Turkish production of Vikhr anti-tank missiles to the offset. After the first 12 helicopters were delivered from Russia, another 38 were to be produced in Turkey. At the same time, Kamov offered a choice of three variants of the helicopter: the single-seat Ka-50 and the two-seat Ka-50-2 and Ka-52. Later, the Italian Agusta Westland, Franco-German Eurocopter and South African Denel also joined the competition. Visits by the Turkish military to Russia and Russian officials to Turkey recommenced; Turkish pilots were flying Ka-52 '061' again. Kamov received very strong support from the Russian authorities, including President Vladimir Putin during his visit to Turkey in December 2004.

It did not help much. The Agusta A129 Mangusta and Denel Rooivalk were shortlisted on 22 June 2006, and finally on 30 March 2007, the victory of A129, which was assembled in Turkey by Turkish Aerospace Industries (TAI) as the T129, was announced. On 7 September 2007, a US$1.2 billion contract was signed.

The greatest benefit for Kamov from participating in the Turkish tender was getting to know the 'open architecture' of the on-board equipment. For the first time, the Russians connected the

For export: the first attempt

A mock-up of the final tandem Ka-50-2 version, as presented at the ILA exhibition in Berlin in 2002. In place of the Russian cannon, the helicopter has a French 20mm Giat 621 cannon mounted on a flexible carriage under the fuselage. (Piotr Butowski)

helicopter's equipment through two data buses (one for the pilot-navigation equipment and the other for the weapons control system), made in accordance with the Mil Std 1553B standard, with universal data exchange protocols. This made it much easier to modify and modernise the equipment than in the 'rigid' systems made for specific devices, used before by the Russians.

Even earlier, in Israel, the Russians became acquainted for the first time with modern night-vision goggles (NVG) for flying at night; they had no experience in this field at the time. The problem was not the goggles themselves – these were provided by Israel – but the adaptation of the lighting of the instrument panels in the cockpit to fit them, as well as the technique of piloting the helicopter at night using such equipment. In May 1999, Kamov's test pilots, Alexander Smirnov and Oleg Krivoshein, arrived in Israel. The Israelis provided the Russians with a light helicopter Bell 206 for flights with the NVGs. Kamov gained invaluable experience, which later became useful in making Ka-52 equipment for the Russian Air Force. The first flights in '061' with NVGs at 0.001 lux light were made in June 1999.

Kamov/IAI Ka-50-2 Erdoğan Helicopter Specifications

Principal dimensions:
Rotor diameter: 47ft 4in (14.43m) upper, 47ft 5in (14.46m) lower
Maximum length, rotors turning: 52ft 4.5in (15.96m)
Fuselage length: 46ft 4in (14.13m)
Wingspan: 23ft (7.01m)
Maximum height: 16ft 3in (4.95m)

Undercarriage:
Type: Retractable with twin nose wheels and single main wheels.
Wheel base: 14ft 11in (4.54m)
Wheel track: 8ft 9in (2.67m)

Weights:
Normal take-off: 21,605lb (9,800kg)
Maximum take-off weight: 24,912lb (11,300kg)

Performance:
Maximum level flight speed: 167kts (310km/h)
OGE hover ceiling: 11,810ft (3,600m)
Practical range: 323 miles (520km)
Ferry range: 721 miles (1,160km)

In South Korea

In the spring of 2000, the requirements of a tender for 36 AH-X combat helicopters were announced in Seoul, South Korea. Similar to the Turkish tender, some helicopters were to be delivered by the original manufacturer, with others would be made by Korea Aerospace Industries (KAI). Kamov (of course, through the state intermediary Rosvooruzheniye) came there with a Ka-52K (Korean) proposal. The Russians had a chance, if only because Korea used (and still uses today) several dozen other Kamov helicopters – the civilian Ka-32s. By now, Korea had bought 59 of them (in 2000, the country had about 30), including seven examples for the Republic of Korea Air Force, where they serve as search and rescue helicopters under the local designation HH-32A; the Russian designation for this version is Ka-32A4. Another eight examples are operated by the South Korean Coast Guard.

The peculiarity of the tender in Korea was the ban on direct contact by the bidder with the Korean Ministry of Defence or KAI – this was why a Korean intermediary was needed. The Russians chose the LG corporation, which had previously dealt with the purchase and after-sales service of Ka-32 helicopters. The Mi-28N was also initially submitted to the Korean tender, but it was quickly withdrawn by the Russians.

Unlike with Turkey, a helicopter variant with all-Russian equipment and armament was prepared for Korea; IAI, which Kamov initially offered to cooperate with, presented too many inconvenient requirements. By this time, the Russians had already mastered Mil Std 1553B, and there were several companies in Russia capable of taking on such a task. As the equipment integrator, Kamov chose the Kronshtadt company (formerly Transas), which made the Kabris digital map for the Ka-50.

However, things began to change. The Koreans insisted on removing the systems used in North Korea, for example the 2A42 cannon (used on BMP-2 armoured personnel carriers). Like in the Turkish tender, the Ka-52K helicopter was to be fitted with the French Giat cannon and 70mm NATO unguided rockets. The Russian Vikhr ATGM was to remain the main anti-tank armament, but the Koreans also asked them to consider the Israeli NT-D missile. The French Thales Topowl helmet-mounted display and target indicator was planned for the Ka-52K. Of course, the Russian self-defence system was to be replaced by BAE Systems' suite, and the encrypted communication by Marconi was requested. The identification friend or foe (IFF) system was supposed to be Korean.

As usual, there were mutual visits. Delegations from South Korea visited the Kamov helicopter flight test site at Chkalovsky and the Kronshtadt (avionics) and Krasny Oktyabr (helicopter's gearbox) facilities in St Petersburg. In April 2001, pilots from Korea made familiarisation flights on Ka-52 '061' in Torzhok. However, just before the planned decision date, the South Korean tender was unexpectedly closed, without explanation.

Apart from Turkey and South Korea, talks were underway on the possible sale of the Ka-50s/Ka-52s to Finland and Singapore, as well as China, India, Malaysia, Myanmar and Syria.

Ka-52 '061' in the Ka-52K (Korean) configuration, in which, in 2000–01, it participated in the tender for 36 AH-X combat helicopters for South Korea. The Samshit sighting turret is located in the very front of the fuselage. (Kamov)

Chapter 4

Ka-52 two-seater: the beginning

The fact that the Soviet MoD accepted the single-seat Ka-50 did not mean they gave up on the two-seat option. In 1985, Kamov was designing the V-80UB (Uchebno-boyevoi, combat trainer) two-seater version. At that time, the designers acted in a traditional way – they lengthened the nose of the helicopter and placed a second cockpit there, in front of the regular one. It was then they realised the method that worked for aeroplanes (for example, for the training versions of the MiG-25 or Tu-128 fighters) was not suitable for a helicopter: adding weight to the front significantly changes the position of the centre of gravity and requires significant reworking of the entire airframe. The V-80UB project did not go beyond the front fuselage mock-up stage.

Today's Ka-52 began with the Soviet government's resolution on helicopters on 14 December 1987, which was mentioned previously. Fulfilling the resolution, on 15 June 1990, the USSR MoD ordered a research and development (R&D) work from Kamov, codenamed Avangard-1, to create a two-seat reconnaissance and combat helicopter, the V-80Sh2. A year later, Mil was ordered to produce a night version of the Mi-28N, or Avangard-2. The order of the V-80Sh2 did not mean – back then – the abandonment of the concept of the Ka-50 (V-80Sh1) single-seat combat helicopter. In principle, it was the Ka-50 that was to be the basic combat helicopter of the Soviet Union, and the two-seat Ka-52 was to perform only selected tasks.

This time, when arranging the crew of two, Kamov's designers insisted on an unusual solution: they put the pilots side by side, while in all other combat helicopters in the world, the crew sit in tandem. First of all, such a cockpit required the least changes to the original design of the Ka-50. The front of the fuselage became heavier but also shorter, and it was not necessary to rework the entire airframe.

The front fuselage mock-up of the first V-80UB two-seat training version of 1985; it was abandoned after a short time. (Kamov)

The first, rough, mock-up of the Ka-52 (V-80Sh2) fuselage as we know it today, with two crew members sitting side by side; seen here at MAKS 1995. (Piotr Butowski)

The fuselage was cut at the frame number 7; a new wide cockpit was installed in the front, while the rear part remained unchanged. Note also that the front of the Ka-50's fuselage was much narrower than the mid-section; after replacement, the wider two-seater cockpit did not protrude beyond the fuselage perimeter in plan view. According to Kamov, it was easier for pilots sitting next to each other to cooperate; the controls are the same for both seats, and any crew member can fly the helicopter. Another advantage of the wide front fuselage is that it is easy to place a powerful radar there. There are also disadvantages of this solution – mainly limits to the visibility of the left pilot to their right and vice versa.

The first mock-up of the two-seater V-80Sh2 was presented to the commission in September 1994 and did not arouse enthusiasm, but the idea of placing pilots side by side was accepted. The refined

Above left: The pilots' cockpit in the Ka-52 mock-up shown during MAKS 1995. (Piotr Butowski)

Above right: The first Ka-52, '061', in its initial equipment configuration of 1996 had a large rotating cylinder of the Rotor electro-optical targeting system under the nose and a ball of another Samshit-BM system at the top, behind the cockpit. The helicopter has never flown with this configuration of equipment. It began its flight tests on 25 June 1997 without the Rotor and Samshit devices. (Alexei Mikheyev, Kamov)

Above left: In the wide front of the Ka-52 fuselage, there was room not only for two crew members, but also for a large Arbalet radar. (Alexei Mikheyev, Kamov)

Above right: Ka-52 '061', shown here at MAKS 1997, was created as a conversion of very popular single-seat Ka-50 '021', which appeared many times at international airshows. (Piotr Butowski)

Right: Ka-52 '061' firing the 2A42 cannon on a ground testbed. (Kamov)

design was approved inside the company on 15 March 1995, after which the Kamov workshop in Ukhtomskaya began to convert one of the Ka-50s into a prototype Ka-52. It was '01-02', the second helicopter of the first production series, which previously had the side number '021' and was presented at Le Bourget in 1993. When converted into the first Ka-52, it was given the number '061', which was derived from the internal designation of 'izdeliye 800.06'.

The planned two-seat Ka-52 variant was publicly announced for the first time at Farnborough in September 1994. In Paris in June 1995, Kamov released the first approximate sketch of the Ka-52, while at Zhukovsky in August 1995, a rough black-painted fuselage mock-up was displayed. In early December 1996, at the Aero India exhibition in Bangalore, Kamov brought Ka-52 '061' prototype, which was called 'Alligator'. Single-seat Ka-50 '024' Black Shark also arrived in India.

Ka-52 '061' made its maiden flight on 25 June 1997, piloted by Alexander Smirnov and Dmitry Titov, almost ten years after the government resolution on V-80Sh2. And then, once again, there was an 11-year hiatus: the second, '062', was not made until 2008. This does not mean that nothing happened in the Ka-52 programme – '061' was being tested in constantly changing equipment configurations.

Above: Ka-52 '061' fires a salvo of unguided rockets. (Kamov)

Left: In 2005, Ka-52 '061' flew in this configuration with the Arbalet-52 radar in the front of the fuselage, the Samshit-BM aiming turret on the ridge behind the cockpit and a small observation TOES-520 turret under the front of the fuselage, on the left side. Additionally, a small Arbalet-D radar was mounted on top of the rotor head. The helicopter is, here at MAKS 2005, presented armed with R-73 air-to-air missiles, although they were never integrated with the Ka-52 equipment. (Piotr Butowski)

Above: This is how Ka-52 '061' looked like in 2007. The optoelectronic sensors have been removed, and the perennial black livery has been replaced with a two-colour camouflage pattern. (Kamov)

Left: The L-band (decimetre wavelength) Arbalet-D radar by Phazotron-NIIR company, with the antenna in a small 'egg' placed over the rotors, was used for detecting approaching missiles and aircraft. In the current operational Ka-52 helicopters, the idea remains, but the self-defence L-band radar has four antennas built into the airframe. (Piotr Butowski)

Above left: This is how the cockpit of '061' changed. This is how it looked like when the helicopter began its flight tests in 1997... (Piotr Butowski)

Above right: ... and here, ten years later, in 2007. (Piotr Butowski)

Right: During MAKS 2007 in Zhukovsky, Ka-52 '061' was shown with a pack of four tube-launched Hermes-A guided missiles. It was the only presentation of the Ka-52 with these missiles. (Piotr Butowski)

In the first version, shown in December 1996 in Bangalore, the helicopter had a large rotating cylinder of the rotor EO targeting system under the nose; the rotor, made by Zenit from Krasnogorsk, was a variation of the Tor-28 sensor that is still used today in Mi-28N combat helicopters. The rotor differed from the Ka-50's Shkval (developed by the same Zenit company) with the addition of the thermal imaging camera (it was the French Victor by Thomson-CSF) and larger observation angles (±110°) in azimuth. Directly above the rotor, the Myech-U (Arbalet-52) radar by Phazotron-NIIR of Moscow was to be installed (in fact, it was not there). On the fuselage top, just behind the crew cockpit, there was a ball of a large EO aiming system Samshit-BM, made by UOMZ from Yekaterinburg. Another small observation turret was placed on the underside of the fuselage. It is clear that the number of sensors placed on '061' at that time were redundant and more of a demonstration of the possibilities, rather than a realistic set of equipment.

During the next presentation at Zhukovsky in August 1997, the set of equipment was more modest; only the Arbalet radar in the front fuselage and the Samshit ball on the back remained. For a short time, to meet the requirements of the tender in South Korea, '061' was converted into the Ka-52K version. It had no radar and only two EO turrets – a large one for the WSO in the nose and a small one for the pilot underneath. During the presentation of the helicopter for the tender in Turkey, '061' received the inscription 'Ka-50-2' on the tailfin, as it was entered into the competition with such a designation. In Turkey, the helicopter was to show only its flight handlings, and all sensors were removed for that time.

In the following years, the helicopter flew with the aiming Samshit-BM turret on the ridge behind the cockpit and a small observation TOES-520 turret under the front of the fuselage, on the left side. In 2001–05, Ka-52 '061' was presented with an Arbalet-D radar with the antenna in a small 'egg' placed

The Hermes-A missile (this is its export designation; the internal Russian name is Klevok, peck) has a two-stage rocket motor, enabling a maximum speed of 1,000m/s and a range of 7.5–9.3 miles (12–15km). (KBP)

over the rotors. The Arbalet-D worked in the decimetre range (L-band) and was used for self-defence – detecting approaching missiles and aircraft. In the current operational Ka-52 helicopters, this idea remains, but the way it is implemented is different: the decimetre-wave radar for self-defence has four antennas built into the airframe, providing 360° coverage.

'061', like the Ka-50 before it, was presented several times armed with the Kh-25M air-to-ground guided missiles, as well as the R-73 air-to-air missiles, although they were never integrated with the Ka-52 equipment.

During MAKS 2007 in Zhukovsky, Ka-52 '061' was shown with a pack of four Hermes-A large guided missiles; this is the export designation of the Klevok-V ('peck', like a bird) missile. The Hermes by the KBP design bureau is a much larger development of the Vikhr missile, with a two-stage rocket motor enabling a maximum speed of 1,000m/s; the missile length is 11ft ½in (3.5m) and its weight is 243lb (110kg), including a 62lb (28kg) warhead. The Klevok-V (Hermes-A) helicopter version has a range of 7.5–9.3 miles (12–15km) with the possibility to extend this to 12.4 miles (20km) when radio mid-course correction is added to the basic inertial navigation. A small semi-active laser seeker manages terminal guidance, with two channels enabling the simultaneous launching and homing of two missiles. Other terminal seekers, including passive infrared and active radar, are being considered. It was the only presentation of the Ka-52 with these missiles ever, although KBP still advertises the Hermes-A missile as a possible weapon for the Ka-52.

Step by step, though still very slowly, the equipment for the Ka-52 became real, not just mock-up versions. In March 2004, the helicopter began trials with an Arbalet-52 radar fitted in the forward fuselage. In December 2005, a Ka-52 fired a 9M120-1 Ataka-1 missile for the first time.

Above left: In 2009, Ka-52 '061' has an undercut front of the fuselage and a bullet imitating the Samshit under it. A package with six Ataka anti-tank missiles hangs under the wing. The debris is not a dirty lens on the camera, but grass leaves raised by a stream of air from the rotors. (Piotr Butowski)

Above right: Ka-52 '061' at the MAKS exhibition in August 2009 in a joint flight with its largest competitor, the Mi-28. This '35' is one of the Mi-28N helicopters of the initial series, used for tests by the Mil team. (Piotr Butowski)

Chapter 5
Ka-52 in production and service

The 1990s and early 2000s were a very difficult period in Russian aviation. Russia faced economic turbulence and cut their military spending drastically. Deputy commander-in-chief of the Russian Air Force for procurement, General Major Dmitry Morozov, complained in August 2003 that only 50 per cent of the aircraft in the air force were airworthy. Deliveries of new aircraft had essentially halted almost ten years earlier. The rebound began after 2000, when a long-term rise in the price of crude oil, the main driver of the Russian economy, began on the world markets; starting at US$20 a barrel, in February 2008 it exceeded US$100 for the first time in history. After a ten-year hiatus, the Russians began refreshing their air force equipment; in 2005, the MoD placed the first orders for new aircraft and helicopters, including 67 Mi-28Ns. In the following years, orders took off exponentially.

The Ka-52 programme has also been restarted. At the end of 2006, the MoD ordered Kamov to complete development of the helicopter in compliance with updated requirements. Following this, on 13 March 2007, the Progress plant in Arsenyev received an order to produce a preliminary batch of six helicopters for trials at Kamov, with delivery by 2010. The first two test specimens, '062' and '063', were made of parts previously produced for the Ka-50. '062' made its first flight on 26 June 2008, followed by '063' on 28 October. In December 2008, formally belonging to Kamov, both helicopters arrived at Torzhok, where they joined the trials. Next, three pre-series helicopters ('51', '52', and '53') were also handed over to Kamov in 2009; the sixth helicopter of the March 2007 order is unknown.

From there, progress increased rapidly. In 2010–11, the military centre in Torzhok received nine Ka-52s with numbers from '91' to '99'; the first four arrived in December 2010 and began operations at Torzhok on 8 February 2011. The previous helicopters, although operated in Torzhok, were owned by Kamov. The Ka-52 officially completed the state evaluation on 20 November 2011.

In 2013, Ka-52 '062' was still flying with a ball imitating the GOES optoelectronic turret, but it had already received part of a real self-defence system, including L370-2 warning sensors and L370-5 jammers. (Piotr Butowski)

This '063' is the third experimental Ka-52; it began flight tests on 28 October 2008. (Piotr Butowski)

The first production helicopters were incomplete. For example, helicopters '51' and '52' had mock-ups instead of real GOES-451 aiming payloads and no self-defence or radar systems at all. In the next helicopters, '53' and '91' to '99', the optical systems were there, but the radars were still absent. The first helicopters equipped with radars were delivered at the end of 2011. The previous ones were retrofitted with radars later.

For years, the Ka-52 was presented as a niche product, which was to be used in the Russian armed forces for unspecified 'special tasks'. Back in January 2003, the then commander-in-chief of the Russian Air Force, General Vladimir Mikhailov, declared the service had chosen the Mi-28N as its future basic combat helicopter, while the Ka-52 would be produced in a limited number and 'will find a place in special forces, where some of its unique characteristics will be useful'. Kamov's head, Sergey Mikheyev, replied that the Ka-52 is the only combat helicopter in Russia that truly works in all conditions, not only day and night but also in fog, rain, smoke screen and other obstacles, while the Mi-28N 'will require many more tests to reach the level already achieved by the Ka-52'.

There seemed to be many indications that the MoD intended to close the Ka-50/Ka-52 programme. Sergey Mikheyev then complained that in 2004–05 the Ka-52 programme only received about 1 per cent of the funds allocated to the Mi-28N from the state budget at the same time. Kamov was looking for its niche, emphasising that the helicopter is irreplaceable on operations in the mountains (like in Chechnya), where its greatest advantages are a high-hover altitude and exceptional manoeuvrability. Colonel Alexei Lande, head of combat preparation for transport and special military helicopters, told the *Krasnaya Zvezda* newspaper that the Ka-50 could 'hover at 4,000m above sea level and fire at targets from there, which no other Russian helicopter could do'. Therefore, 'it would be advisable to create a sub-unit of such helicopters to perform special tasks in the North Caucasus'. However, the praise did not seem to help much.

It was a big surprise then when the Russian MoD signed a contract on 1 March 2011, ordering 146 Ka-52 helicopters for delivery by 2020. Thus, the Ka-52 transformed from a niche 'command' helicopter into an alternative to the Mi-28N attack helicopter; the number of ordered helicopters of both types is roughly the same. According to Russian Helicopters, the contract for the 146 Ka-52s was worth 120 billion roubles, which means the price of one helicopter was around US$25 million at the then exchange rate.

On 23 May 2011, deliveries of the Ka-52 to the first operational unit commenced, the helicopters going to Chernigovka (37 miles [60km] from the Arsenyev production plant). In February 2013, the first Ka-52 helicopters entered service at Korenovsk near the Caucasus; in 2014, they arrived in Ostrov,

Ka-52 production line at the Progress plant in Arsenyev in the far east of Russia. (Russian Helicopters)

Khabarovsk and Dzhankoy (in the Crimea), followed by Vyazma in 2018 and Zernograd in 2020. In 2021, four Ka-52s were deployed to the Russian base in Erebuni, Armenia.

In January 2022, the commander of the Central Military District, Alexander Lapin, announced that by the end of the year, a squadron of Ka-52 helicopters would be assigned to an army aviation brigade (17 Br AA) in Kamensk Uralsky.

The Russian Air Force has already received all 146 Ka-52s ordered in 2011. The production rate at Arsenyev is 15–20 helicopters per year; 14 helicopters were delivered in 2021. The order for 30 Ka-52M helicopters placed in August 2021, and the promised contract for another 84 helicopters, means that the Russian military will continue to receive helicopters at a similar pace until 2027.

How to read the construction number of a Ka-52

The Ka-52 construction number is a sequence of 11 digits (until 2012, it was 13 digits). The number is marked on a helicopter in several places, for example at the bottom of the tailfin, underneath the tail plane and on the side of the aft part of the fuselage.

The first three digits are always '353' – this is the code of the Progress production plant in Arsenyev. The exceptions are the first test Ka-50 helicopters, and Ka-52 '061', which have the first three digits of '879', the code of the Kamov workshops in Lyubertsy.

The next three digits identify the version of the helicopter:

800 – is the test version of the Ka-50 helicopter; Ka-52 '061' also has this number.
805 – is a series production Ka-50 helicopter, or 'izdeliye 800.05'.
826 – is a series production Ka-52 helicopter for the Russian Air Force, or 'izdeliye 800.06'.
820 – is a Ka-52K ship-based helicopter, or 'izdeliye 800.20'.
830 – is an export Ka-52E helicopter, or 'izdeliye 800.30'.
850 – is a modernised Ka-52M helicopter, or 'izdeliye 800.50'.

The next two digits are the production batch number, and the last three digits are the helicopter number in the batch, from '001' to '010'. From the eighth production series (2013), there are ten helicopters in each batch; before that, the production batches had five helicopters. In 2021, the 18th and 19th batches were in production.

As an example, the helicopter shown in the top two photos on page 48, presented at the Army 2019 forum in Kubinka, has construction number 35382615009. We can see then that it was

produced in Arsenyev (353), was the standard Ka-52 (826), was made in the 15th production batch (15) and was the ninth helicopter (009).

Russian military aircraft have two-digit tactical numbers (in training units sometimes three-digit), usually in red or blue with a thin white outline; this sample helicopter has a red '87' number with a black outline. Yellow or white numbers are rare. As a rule, aircraft of one squadron wear numbers of the same colour; occasionally, when an aircraft is moved from one unit to another, for some time, it may keep its 'old' colours. There are no strict rules to assigning these numbers. Usually, brand-new aircraft arrive from the production facility with successive numbers, but not necessarily. Known examples include one squadron with odd and the other with even numbers. One should not attach too much importance to these numbers, since they may be changed. For instance, after their return from Syria, the markings on many aircraft were revised.

The aircraft of the MoD (as well as other governmental structures like Rosgvardiya, Federal Security Service [FSB] and Federal Guard Service [FSO]) also have registrations consisting of the letters 'RF' and five digits; in this case, it is RF-13426.

Until the end of 2012, Ka-50 and early-production Ka-52 helicopters had serial numbers consisting of 13 digits. The first six digits denoting the manufacturer and version of the helicopter, and the last five digits representing the production batch number and the helicopter number in the batch, had the same meaning as described above. However, two additional digits were inserted between them, indicating the quarter and year of production. For example, the Ka-52 helicopter pictured in the bottom two photos with the yellow side number '97' and registration RF-91264, has the construction number '3538264105002'. This means it is the second helicopter of the fifth production batch made in the first quarter of 2014.

Above left: Ka-52 '51' is the first helicopter of the first production series, built in the third quarter of 2009 and delivered to the Kamov company for continued testing. (Piotr Butowski)

Above right: Ka-52 '52' is the second helicopter of the first production series. Like '51', it only has mock-ups of the equipment. The helicopter crashed on 29 October 2013 during tests at the Kamov airfield. (Piotr Butowski)

Below left: The third initial-series helicopter, '53', had already received the GOES-451 optoelectronic sighting turret but still lacked the self-defence system; there are only empty stands for the devices. (Piotr Butowski)

Below right: Ka-52 '93', c/n 3538264003003, is one of the first serial helicopters, manufactured in the fourth quarter of 2010, which is used in the crew conversion centre at Torzhok. (Piotr Butowski)

Ka-52 operational units in the Russian Aerospace Forces

Army aviation in Russia was born on 28 October 1948. On this day in Serpukhov, near Moscow, the first training-liaison squadron was created with G-3 helicopters by Ivan Bratukhin. The first regiments of transport helicopters were formed in the mid-1950s, along with the start of series production of the Mi-4 helicopter. The first armed helicopters, Mi-4AVs (converted from transport Mi-4As), entered service near the end of the 1960s; combat Mi-24s entered service in 1971.

In the USSR (and now Russia), there has always been a discussion about the right place for helicopters – with the ground troops (army) or in the air force? Before 1990, the helicopters were subordinated to the air force. In 1990, as a result of the Afghanistan War (1979–89), they were incorporated into the ground troops' structure. On 1 January 2003, combat and transport helicopters returned to the air force; Russian ground troops have had no air assets since then. The decision to

Locations of Russian Aerospace Forces units operating Ka-52 helicopters. (Piotr Butowski)

transfer the helicopters back to the air force in 2003 was made following the shock shooting down of a Mi-26 transport helicopter in Chechnya and the death of 119 people on board on 19 August 2002. The investigation revealed negligence and numerous weaknesses of the army aviation command system within the ground troops.

Military helicopters in Russia are organised into independent helicopter regiments (Otdelnyi Vertolyotnyi Polk, OVP) and army aviation brigades (Brigada Armeyskoi Aviatsii, Br AA). The helicopter regiment has, in accordance with complement, three squadrons of 12–16 helicopters each and a couple of command or auxiliary helicopters. The regiment's line-up is mixed: two combat helicopter (Ka-52, Mi-28N, Mi-24/Mi-35M) squadrons and one transport helicopter (Mi-8) squadron or, inversely, one combat squadron and two transport squadrons. The brigade is bigger than the regiment: it has two squadrons of attack helicopters and two squadrons of transport helicopters, plus some heavy-lift Mi-26s.

On 23 May 2011, deliveries of Ka-52s to the first operational unit in Chernigovka commenced. (Russian Ministry of Defence)

Helicopter units are subordinated in two ways: they are operationally managed by the joint strategic commands of individual military districts, while training and combat preparation is handled by Aerospace Forces. Aerospace Forces' headquarters are in Moscow, with a military aviation combat preparation department headed by Major General Oleg Chesnokov. Russia is divided into five military districts (MD): Western MD, Southern MD, Central MD, Eastern MD, and the fifth district is the Northern Fleet. In each of the districts, except the Northern Fleet, there is one army aviation brigade and one or two helicopter regiments. Ka-52 helicopters are currently in service with three brigades and four regiments; they are also at the military aviation's crew conversion centre in Torzhok.

344th State Combat Training and Flight Crew Conversion Centre of Army Aviation
HQ: Torzhok
The 344th Centre is tasked with conducting military evaluations of helicopters and the conversion of pilots to new types. Torzhok always receives new helicopters first; it has operated the Mi-28N since 2008, the Ka-52 since December 2010 and the Mi-35M since 2011. Since 12 April 1992, the Centre has included the Berkuty (Golden Eagles) aerobatics team, flying Mi-28N helicopters (Mi-24s until 2011). The helicopters of the 344th Centre are grouped within two research-instructor units. Most are assigned to the 696th Research Instructor Helicopter Regiment at Torzhok, while several Mi-8s and Mi-24s are operated by the 92nd Research Instructor Helicopter Squadron at Klin airfield, 130km away from Torzhok, towards Moscow.

Western Military District
15th Army Aviation Brigade
Location: Ostrov
This is a totally new formation without any historical traditions. It was formed in 2013 at Ostrov Air Base, where the naval aviation's crew conversion centre was based until 2009. Since mid-2013, the brigade has been receiving Mi-28Ns and Mi-35Ms, and Ka-52 helicopters since 2014. Currently the brigade has a squadron of 12 Ka-52 attack helicopters, a squadron of 12 Mi-28Ns and four Mi-35M attack helicopters, two squadrons of Mi-8MTV-5 assault helicopters and a detachment of four Mi-26 heavy-transport helicopters. The brigade also operates Mi-8MTPR-1 Rychag electronic warfare helicopters.

Above left: Ka-52 '23', RF-91123, has served in the 318th Independent Helicopter Regiment in Chernigovka since 2013. (Russian Ministry of Defence)

Above right: The Ka-52 helicopter cockpit simulator used to train crews at the Chernigovka base. (Russian Ministry of Defence)

440th Independent Helicopter Regiment
Location: Vyazma
The 440th regiment with Mi-8 and Mi-24 helicopters (including K and R special purpose versions) was formed in 1987 in East Germany. In July 1992, the regiment was withdrawn to Vyazma, where the Germans built a housing facility for the unit's personnel. Until 2014, the unit operated old Mi-24 and Mi-8 versions, with one squadron of each type. The re-equipment began in 2014, when the Mi-28Ns and then Mi-8MTV-5s arrived. In March 2018, the regiment received the initial eight Ka-52 helicopters. Currently, the regiment has one squadron of Ka-52 helicopters, a squadron of Mi-24 and Mi-28N helicopters and a squadron of Mi-8s (including the Mi-8MTPR, Mi-8SMV and other electronic countermeasures [ECM] versions). Several dozen unairworthy Mi-8s and Mi-24s are standing at the periphery of the airfield. The Rus aerobatic team comprising L-39C aircraft is based at the same Vyazma airfield.

Southern Military District
16th Army Aviation Brigade
Location: Zernograd
The brigade was formed on 1 December 2015 at an airfield in Zernograd, which had been non-operational since 2009. The core of the new brigade was provided by the 546 Aviation Base moved from Rostov-on-Don. The Ka-52 squadron was fully completed in December 2021. The brigade also has one squadron of Mi-35M and Mi-28N attack helicopters and two squadrons of Mi-8 and Mi-26 transport helicopters (including some Mi-8 EW versions).

39th Helicopter Regiment
Location: Dzhankoy
The regiment was formed on 3 July 2014, shortly after Russia occupied Crimea. The first squadron, with Ka-52 helicopters, arrived from 55th Regiment based at Korenovsk; later, the second squadron, equipped with Mi-28N and Mi-35M combat helicopters, and the third squadron on Mi-8AMTSh transport helicopters, were formed; the squadrons are not complete.

Above left: Ceremonial commissioning of the first Ka-52s at the base in Korenovsk, 18 February 2013. After the seizure of Crimea by Russia in 2014, the Ka-52 squadron from Korenovsk was transferred to the Crimea and became the nucleus of a new helicopter regiment in Dzhankoy. (Russian Ministry of Defence)

Above right: The Korenovsk-based 55th Independent Helicopter Regiment obtained a new Ka-52 squadron, including '90' RF-13429, in 2016–17. (Piotr Butowski)

55th Independent Helicopter Regiment
Location: Korenovsk

The unit's history dates back to 1942; since 1992, it has been based at Korenovsk. The regiment's crews took part in all conflicts the USSR and Russia have been involved in, including Afghanistan, Chechnya, Dagestan, Georgia and Syria. According to the official history, 'The regiment actively participated in the returning of Crimea into the Russian Federation'. On 8 July 2016, Colonel Ryafagat Khabibullin, the then commander of the regiment, was killed in Syria in a Mi-35M. The regiment has the title Sevastopol and Kutuzov order; the regimental day is 20 March.

The 55th Independent Helicopter Regiment currently has a squadron of Ka-52 attack helicopters, a squadron of Mi-28N (October 2010) and Mi-35M (2012) attack helicopters and a squadron of Mi-8AMTSh transport helicopters (the first ones arrived in 2010). The Ka-52 squadron was formed twice – the first time in 2013. The helicopters were delivered to Korenovsk by the Rostvertol plant in Rostov. First, the Ka-52s were airlifted in a disassembled state to Rostov. There they were assembled by a team from Arsenyev, and then they flew to Korenovsk (about 200km south of Rostov). The first Ka-52 arrived at Korenovsk on 18 February 2013. After the seizure of Crimea in 2014, the squadron was moved to Dzhankoy to form the new 39th Regiment around it. The Korenovsk regiment obtained a new Ka-52 squadron in 2016. During 2013–16, major reconstruction of the airfield was done – in fact,

Right: Ka-52 '44' RF-91336 from the helicopter brigade in Ostrov. Note the GOES-451 turret in its working position. (Piotr Butowski)

Below: Ka-52 '45' RF-91337 from the brigade in Ostrov is towed to its parking position after completed flights. (Piotr Butowski)

it was built from scratch. The concrete runway (8,530ft x 138ft [2,600m x 42m]) was built (previously there was a grass strip), as well as canopied parking places for helicopters, which is unique in Russia.

3624th Aviation Base
Location: Erebuni, Armenia
Russia maintains the 102nd Military Base in the town of Gyumri in Armenian territory; the base is subordinated to the Southern MD. In accordance with an agreement of 2010, Russia is entitled to retain forces here until 2044. The air component at Erebuni has one squadron of MiG-29 and Su-30SM fighters and one of Mi-8, Mi-24 and Ka-52 helicopters.

Central Military District
17th Brigade of Army Aviation
Location: Kamensk Uralsky
The airfield at Kamensk Uralsky (also called Travyany after a neighbouring housing estate) had been vacant since 1998. The airfield was revived in December 2011, where the 48th Aviation Base was formed by gathering helicopters from disbanded units. On 1 December 2018, the base was transformed into the 17th Army Aviation Brigade. In 2022, the formation of the Ka-52 helicopter squadron within the brigade began. Besides that, the 17th Brigade has one Mi-8MTV-5 squadron and one Mi-24P squadron at Kamensk Uralsky, and a squadron with Mi-8 and Mi-26 helicopters at Uprun (Yuzhnouralsk) airfield, 143 miles (230km) south of Kamensk.

Eastern Military District
18th Brigade of Army Aviation
Location: Khabarovsk Tsentralnyi
The 18th Brigade was formed on 1 December 2016 from the 573rd Aviation Base of Army Aviation, which in turn was the successor to the 825th Helicopter Regiment (at Garovka, near Khabarovsk) that was disbanded in 2009. In recent years, the unit's equipment has been modernised: in 2013, older Mi-8 versions were replaced by the new Mi-8AMTSh; Ka-52 combat helicopters arrived during 2014–15. The brigade currently has one squadron with 18 Ka-52 attack helicopters, one squadron with 18 Mi-8AMTSh helicopters and one squadron with six heavy-lift Mi-26s and several Mi-8s, including Mi-8SMV electronic warfare versions. The brigade also maintains two duty posts, one on the island of Iturup (Etorofu) in the Kuril Islands and one at Anadyr in the Arctic, each with a detachment of Mi-8AMTSh transport helicopters.

Ka-52 '67' RF-91107 initially served in Korenovsk. In 2014, it was transferred to Dzhankoy in the Crimea. Here, it is waiting for overhaul in 2016. (Piotr Butowski)

Above left: Another Ka-52 ('72', RF-91269) currently serving in the 39th Helicopter Regiment in Dzhankoy (previously in Korenovsk). (Piotr Butowski)

Above right: Three Ka-52s from the 440th Independent Helicopter Regiment in Vyazma. The first Ka-52s were delivered to the base in March 2018. (Russian Helicopters)

Below left: The Ka-52 squadron within the 16th Army Aviation Brigade in Zernograd in the Southern Military District was fully completed in December 2021. (Russian Ministry of Defence)

Below right: Preparation of Ka-52 helicopters for flight in the Zernograd brigade in January 2022. (Russian Ministry of Defence)

319th Independent Helicopter Regiment
Location: Chernigovka
The first batch of Ka-52 helicopters arrived in Chernigovka on 23 May 2011. Chernigovka is located 37 miles (60km) from Arsenyev, where the helicopters are produced. This regiment is considered the successor to one of the first aviation groups established by Vladimir Lenin in 1918, and during the Soviet era it even was named after Lenin. In 1960, the unit was transformed into a helicopter regiment. In 1971, it was the first unit to re-equip with Mi-24 attack helicopters (Mi-24s were manufactured in Arsenyev). The regiment currently has a squadron of Ka-52 attack helicopters, a squadron of Mi-8AMTSh transport helicopters and a handful of Mi-26 heavy-lift helicopters. The regimental day is 5 December.

In Syria

In March 2016, several, possibly four, Ka-52s arrived at the Khmeimim Air Base in Syria, where the Russian air contingent has been deployed since September 2015; later, the helicopters and their crews

Ka-52 from the 440th Independent Helicopter Regiment in Vyazma when deployed to Syria in 2017. (Archives)

exchanged with replacements. The main task of the Ka-52 in Syria was to protect the base itself, as well as to accompany ground transport convoys. On 6 May 2018, one Ka-52 was lost near the city of Mayadin in Deir ez-Zor Governorate due to a flying error; both pilots were killed. Especially for Syria, the GOES-451 EO payload was modernised to increase the target detection and tracking range, and the FH01 radar software was updated. At the Army 2019 exhibition in Kubinka, there was Ka-52 '87', RF-13426, advertised as 'modernized based on the experience of the Syrian campaign', but from the outside, it did not differ from a standard Ka-52.

In Ukraine

On 24 February 2022, Russia invaded Ukraine. Russian Ka-52 combat helicopters have been taking part in the fighting from the very first moments of this war.

Even before the war began, Russia installed large forces in the vicinity of Ukraine, including Ka-52 helicopters from units in Dzhankoy, Zernograd, Korenovsk and Vyazma, as well as from the army aviation brigade in Khabarovsk. Several dozen helicopters, including a Ka-52 squadron, have been located at Donuzlav airfield in Crimea, which has been inactive for many years. A large group was moved to Belarus; squadrons of Ka-52 helicopters (as well as other types) were deployed at the Zyabrauka and Mazyr airfields, located 20 miles (30km) and 30 miles (50km), respectively, from the border with Ukraine.

On the morning of 24 February, the Russians dropped off a large heliborne assault unit at the Hostomel airfield, west of Kyiv, whose task was to take over the government district of the capital and capture the state leadership of Ukraine. Hostomel is a civil airport that operates freight with Antonov Airlines. It was planned that first the helicopters would seize the Hostomel airfield, and soon after that large Il-76 transport aircraft with main forces and heavy equipment would land there.

Several dozen Mi-8 transport helicopters supported by combat Ka-52s and Mi-35Ms took off from the territory of Belarus (from where Hostomel is about 50 miles [80km] away) and flew from the north along the Dnieper River backwaters. About 12 miles (20km) before the target, near Vyshgorod, they were fired on by Ukrainian man-portable anti-aircraft missiles. The helicopters immediately fired their decoys and ran away at low altitude. The helicopter assault did not manage to take over the airport, and the arrival of Il-76 transport aircraft with soldiers of the Airborne Forces was cancelled that day. The fight for Hostomel continued for a few more days.

The helicopters involved in the attack had national markings and numbers painted over; instead, a large white letter 'V' was painted on the engine cover. The ground vehicles of Russian troops attacking

Ukraine from the Belarusian side were also marked with the same sign; equipment coming from the territory of Russia has 'Z' signs, and from the side of Crimea 'O'. These are markings enabling quick identification 'friend or foe'.

The available footage shows that the Ka-52s are most often flown over Ukraine armed with one unguided rocket launcher and a package of 4-5 Vikhr-1 ATGMs. Usually, they also have two additional fuel tanks suspended. As of this writing, in the second half of March 2022, the war is still raging. So far, there is confirmed data on eight destroyed Ka-52 helicopters of the Russian Aerospace Forces; it is possible that there were more losses.

Two Ka-52s were damaged and forced to land during the raid to the Hostomel airport on the first day of the war; their registrations are unknown. In a similar way, the Russians lost another Ka-52 a few days later, on 1 March, near the village of Babyntsi, 20 miles (30km) from Kyiv. This helicopter was from Khabarovsk unit and probably had the '17' blue number and registration RF-90680. The fourth confirmed loss is a helicopter that crashed on 8 March in Zernograd, returning to its air base. The fifth Ka-52 was shot down on 12 March near Kherson. It was '74' red, RF-13409, from the Korenovsk regiment. Pictures of another completely destroyed Ka-52 from Korenovsk, number '76' red and registration RF-13411, appeared on 16 March 2022. Ukrainian forces report a few more Ka-52s being shot down, but this is unconfirmed.

Many helicopters were destroyed at the Russian-occupied Kherson airport as a result of artillery fire carried out by the Ukrainian army in mid-March 2022. In most cases, the type of helicopter cannot be determined from the available satellite photos. Among those that can be recognized are two Ka-52s.

Above left: A Ka-52 forced to land after damage during the assault on Hostomel on 24 February 2022. The helicopter has its national markings painted over; instead, a large white letter 'V' is painted on the engine cover. Dust filters were removed from the air inlets before the action to increase the engines' power. (Archives)

Above right: Local people are watching the Ka-52 from the Khabarovsk brigade, which force-landed on 1 March 2022 near the village of Babyntsi in Ukraine. (Archives)

Right: Burning wreckage of Ka-52 '74' red RF-13409 helicopter from the Korenovsk regiment shot down on 12 March 2022, near the city of Kherson in Ukraine. (Archives)

Chapter 6
Ka-52 in detail

Airframe and systems

The main structural element of the Ka-52 fuselage is a composite glass-fibre box beam with a width and height of 3ft 3in (1m). This box takes the main loads from the rotors, wings, weapons and the main landing-gear legs. Inside this structure are fuel tanks and cannon ammunition. Attached outside, on both sides, are equipment modules covered by external panels that form the external outline of the airframe. This is a reverse order compared to classic constructions, where the force carrying structure is external and determines the external shape of the airframe. The Ka-52 configuration, although heavier than the classic one, has advantages that prevailed: it enables better use of the internal volume of the airframe (to obtain its smaller dimensions) and also facilitates access to the equipment during maintenance. The outer panels have a large surface area and can be easily removed. Construction materials are mainly aluminium alloys; 36 per cent (by weight) comprises composite materials.

The engines, gearbox and rotors are mounted at the top of the box, while the main landing gear is mounted at the bottom, with the gun on the right side. The cockpit is attached at the front, with the tail unit at the rear.

The two-spar stub-wings have no high-lift devices; the wings carry armament pylons and sensor/flare fairings at the tips. The slightly swept small tailfin carries a large rudder; the tailplane carries two endplates. The tail unit is not a structural element, and tests proved the helicopter could fly after the tail boom had been lost. The mechanical flight control system uses hydraulic actuators.

Kamov Ka-52E *Hokum B* helicopter cutaway. (Russian Helicopters)

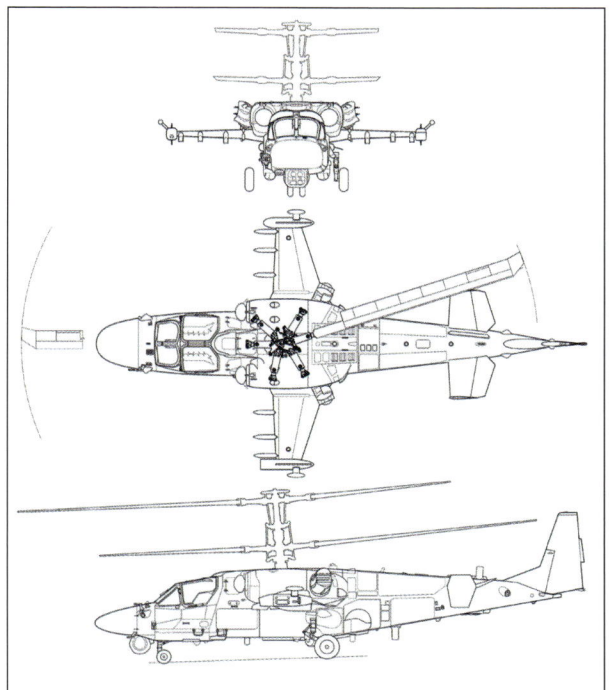

Above left: Three views of Ka-52 *Hokum B*. (Kamov)

Above right: Complex head that bears two three-blade counter-rotating rotors attached by steel torsions. (Piotr Butowski)

The undercarriage comprises retractable twin 15.7in x 5.9in (400mm x 150mm) nose wheels and semi-retractable single 27.6in x 9.8in (700mm x 250mm) mainwheels. In flight, at extremely low altitude, for safety reasons, the landing gear may be extended at any speed. The undercarriage withstands an emergency landing with a vertical speed of 1,970 ft/min (10m/s).

The Ka-52 helicopter systems were made in a manner typical for the previous Kamov helicopters, like the Ka-27 and Ka-32. The hydraulic system is used to drive hydraulic actuators of the flight control system, to retract and extend the landing gear and to move the cannon. The pneumatic system is used to brake the main undercarriage wheels. The electrical system powering the avionics consists of an alternating current 208V/400Hz circuit and a direct current 27V circuit. The sources of AC are two generators driven by the helicopter gearbox. The DC circuit receives its power from the AC system through two rectifiers, as well as from two batteries. The fire protection system is used to extinguish fires in the area of engines, gearbox and fuel tanks. The anti-icing system enables the helicopter to fly in icy conditions. The rotor blades and the pilots' windscreen are electrically heated, the engines air inlets with hot air, and the air-intake dust filters are heated with both hot air and electrically. The air conditioning system is used to maintain the required temperature and pressure in the cockpit. The oxygen equipment is designed to supply the pilots with oxygen when flying above 9,843ft (3,000m).

Powerplant

The Ka-52 is powered by two Klimov/St Petersburg VK-2500 turboshafts. The engine has three power settings, and, at the request of the customer, it can be adjusted to any of them. The most stringent settings are used on the Ka-52, with a take-off power of 1,790kW (2,400shp) and a cruising power

Above left: Main undercarriage leg with a single 27.6in x 9.8in (700mm x 250 mm) wheel. (Piotr Butowski)

Top right: The retractable nose undercarriage has two 15.7in x 5.9in (400mm x 150mm) wheels. The GOES-451 turret is in the cruise position, turned backwards to protect the sensor windows from contamination. (Piotr Butowski)

Above right: The Ka-52 undercarriage in a retracted position. Only the legs of the main landing gear are covered. When flying at extremely low altitudes, the landing gear is usually lowered for safety reasons. (Piotr Butowski)

Above left: Two-stage centrifugal dust filter on the engine air inlet. (Piotr Butowski)

Above right: An infrared suppression cool-air mixer fitted to the engine nozzle. (Piotr Butowski)

of 1,750shp; the specific fuel consumption is 210g/shph. The engine has an emergency power mode, 2,700shp, used in case of failure of one of the engines, which increases flight safety. The VK-2500 engine is 6ft 9in (2,055mm) long, 2ft 2in (660mm) wide, 2ft 5in (728mm) high and has a dry weight of 661lb (300kg). It consists of a two-stage axial compressor, an annular combustion chamber, two two-stage turbines, an exhaust and a drive box for aggregates and systems. Klimov TV3-117VMA engines (1,641kW [2,200shp] each) were used on early helicopters.

Two-stage centrifugal dust filters are installed on the engine air inlets; the degree of air purification is 70–75 per cent. The use of filters significantly increases the engine's service life, but at the same time reduces the power by 2–3 per cent when the filters are off and 4–8 per cent when on. Infrared suppression cool-air mixers can be fitted to the engine nozzles. These are slotted housings in which hot exhaust gas is mixed with the stream of external air. Lowering the exhaust gas temperature reduces the heat trace of the helicopter and thus reduces the likelihood of it being hit by an infrared anti-aircraft missile.

The main engines are started from an Aerosila TA14-130-52 auxiliary power unit (or Ivchenko AI-9K on early helicopters). The TA14-130-52 is a small 105ekW turbine engine located in the upper part of the fuselage, behind the main gearbox. It can be started from the on-board batteries at an altitude of up to 19,685ft (6,000m); on the airfield, it is started from the ground DC power source. The engine is 2ft 10in (868mm) long, 19.1in (485mm) wide, 16.8in (426mm) high and weighs 137lb (62kg). The TA14-130-52 engine also powers the air conditioning system of the cockpit and avionics compartments in flight, as well as diagnostic systems during autonomous operation.

A total of 1,860 litres (409 imp gal, 1,487kg) of fuel is accommodated in two soft tanks within the fuselage. There is also provision for four underwing fuel tanks, each of 540 litres (119 imp gal, 433kg), for ferry flights.

Rotors and transmission

The Ka-52's two lift rotors are co-axial, contra-rotating; the upper rotor turns clockwise (viewed from above) and the lower one turns anti-clockwise. Each rotor has three blades; the upper rotor has a diameter of 47ft 4ins (14.43m), the lower one is 47ft 5ins (14.46m) in diameter. The blades are made

Above left: The Ka-52 can take four underwing fuel tanks, each of 540 litres (119 imp gal), for ferry flights. (Piotr Butowski)

Above right: The Zvezda K-37-800M pilot's seat can be pulled out of the cockpit by means of rocket motor at speeds from 90km/h to 350km/h, and from an altitude of zero to 16,404ft (5,000m). (Piotr Butowski)

Above left: A rocket motor behind the headrest pulls up the cable with the seatback with the pilot attached to it. The seat is pulled slightly to the side, which enables both crew members to be ejected at the same time. (Piotr Butowski)

Above right: Before the seats are fired, the canopy is shattered by this zigzag ribbon of gunpowder and the rotor blades are blown off. (Piotr Butowski)

of composites; they have a speed profile and swept tips. The blades are attached to the rotor head by means of a package of steel torsions that act as a flapping and feathering hinge. The rotors rotate on elastomeric bearings. The rotor hubs and shaft axis are made of titanium.

The VR-80's main transmission gearbox, made by Klimov Company of St Petersburg, has a weight of 2,039lb (925kg); it is driven by two drive shafts which connect it to the engines. Because of widely separated engines, two Klimov PVR-800 intermediate gears have been introduced, each weighting 152lb (69kg).

Crew life support system

The two pilots are seated side by side in a pressurised cockpit, with the commander pilot on the left and the pilot operator on the right. They have the Zvezda/Tomilino K-37-800M ejection (pulled-out) seats that can be operated at speeds from 56mph (90km/h) to 217mph (350km/h), and from zero to 16,404ft (5,000m) altitude. Each seat weighs 121lb (55kg) and is adapted to withstand and absorb high overloads during a hard landing.

When the pilot pulls the handles between their knees, it activates a sequence of actions. First, the belts holding the pilot are pulled in. The rotor blades are detached by means of explosive charges installed in the blade fastenings, and the cockpit canopy doors are blown off. Then the rocket motor located at the rear above the headrest is started. This motor pulls up the cable with the seatback and the pilot attached to it. The two seats are pulled slightly to the side, which enables both crew members to be ejected at the same time. After the used rocket engine separates, the automatic unit unfastens the pilot's straps. The pilot separates from the seat, and the single-dome rescue parachute opens. The pilot has a NAZ-IR rescue vest, which houses a pistol with ammunition, an R-855 radio-beacon, a first aid kit, as well as an ASP-74 rescue belt, which can be refilled with air in water.

Ka-52 pilots have ZSh-7V or VS (Zashchinyi Shlem, crash helmet) helmets by Zvezda. One helmet weighs 3.5lb (1.6kg) and can be used up to a speed of 249mph (400km/h). The pilots have KM-37 oxygen masks connected to the KKO-VK-LP oxygen system (KKO from Komplekt Kislorodnogo Oborudovaniya, oxygen equipment system).

Avionics suite

The BREO-52 ('izdeliye K-806'; BREO means Bortovoye Radio Elektronnoye Oborudovaniye, on-board radio-electronic equipment) integrated avionics suite for Ka-52 has been developed by the Ramenskoye RPKB. The system is controlled by two Baget-53-15-06 computers and integrates targeting sensors, the KBO-806 flight navigation system, the S-403-1 communication system, L370P2 Vitebsk self-defence systems, the Ekran-30-52 built-in checking and recording system and the SUO-806P stores management system.

The KBO-806 (Argument-2000) flight navigation system comprises the SAU-800 autopilot by KBPA of Saratov, INS-2000-02 inertial navigation (error no more than 400m per hour), an A-737-D satellite navigation receiver, a DISS-32-28 Doppler radar, an RSBN-86V TACAN, an ARK-25 direction finder, an A-052-04 radio altimeter, an SVS-V28 air data system, an SIVPV-52 air pressure receiver and others.

The data presentation subsystem utilises four MFI-10-7V displays and two MFPU-01 control consoles installed symmetrically on the wide instrument panel common for both pilots. Additionally, the ILS-28K HUD is located at the left (commander's) seat. The crew is provided with GEO-ONV1-01K NVGs. The S-403-1 communications suite, made by NPP Polyot of Nizhny Novgorod, comprises two UHF radios, an HF radio, an encrypted communication radio and other items.

Targeting sensors

The helicopter has two targeting sensors: the FH01 radar and GOES-451 EO turret. The FH01 Myech-1U (sword) radar suite is made by Phazotron-NIIR of Moscow. Its export derivative, which carries the designation FH01E Arbalet-52 (crossbow, which is an allusion to the American Longbow on the Apache), differs from the Russian version only in its operating frequencies.

The FH01 radar system has two independent radars, the Ka-band and the L-band. The Ka-band (wavelength 8mm) radar has an 800mm-wide mechanical array in the helicopter's nose; the use of such a big antenna is made possible by the wide side-by-side cockpit that creates considerable free space in the helicopter's nose. It is used to detect a target and determine its coordinates. It can select and recognise ground, surface and air targets and then transfers the target data to the EO turret; other tasks include navigation and reconnaissance. The search range for a large ground target (like a railway

Above left: The Ka-band (wavelength 8mm) FH01 radar has an 800mm wide mechanical array in the helicopter's nose. Its search range for a tank is 7.5 miles (12km). (Archives)

Above right: The GOES-451 Samshit gyrostabilised EO turret houses a thermal imaging camera, TV camera, laser rangefinder/designator, laser spot tracker as well as a laser-beam riding system for ATGMs. (Piotr Butowski)

bridge) is 15.5 miles (25km) or 7.5 miles (12km) for a tank. The radar also includes the 423D2 (or 423D12 for export) interrogator of the 40D IFF system.

The weakness of the FH01 radar is the inability of direct guiding anti-tank missiles. For weapon guidance, the target information must be transmitted by the radar to the GOES-451 EO aiming turret, which is the Ka-52's main sighting sensor.

Another radar in the FH01 suite is the N035 (export designation FA01 or Arbalet-D) of the L-band (wavelength: 10cm), which works with the helicopter's self-defence system. The N035 is a warning radar capable of detecting and tracking small air targets, including anti-aircraft missiles, and then transmitting information about the detected threat to the on-board computer.

In the version for Ka-52, the N035 radar has four antennas: two small (32cm x 27cm) flat N035-01 antennas placed on the tailplane endplates and looking to the sides, one dipole N035-01K antenna located in the fairing at the left wingtip, looking forward, and another N035-01K antenna located under the tail boom, looking backwards. Together, they cover 360° in azimuth and -45/+15° in the elevation angle. The manufacturer declares that the FA01 radar can detect a Stinger-class missile from a distance of 1.9 miles (3km) and a fighter aircraft from 6.2 miles (10km).

The GOES-451 Samshit gyrostabilised EO turret produced by UOMZ of Yekaternburg is mounted under the nose. It is used to search, detect and recognise ground and low-speed air targets, then track them and, if necessary, illuminate them with a laser. The turret allows aiming of the cannon and unguided rockets, as well as guiding the ATGMs in the laser beam.

The GOES-451 houses five sensors: thermal imaging (8–12 micrometre) camera, TV camera, laser rangefinder/designator, laser spot tracker and the LSN (Lazernaya Sistema Navedeniya) laser-beam riding system for Ataka-1 and Wikhr-1 ATGMs. The automat, or operator in manual mode, keeps the laser beam on the target while the missile control system keeps it in the centre of the beam.

The diameter of the GOES-451 ball is 720mm and the weight is 485lb (220kg), of which 143lb (65kg) are sensors; the platform with sensors is gyroscopically stabilised in three axes. In the working position, the turret rotates within +/-135° in azimuth and -80/+20° in the elevation angle. In the cruise position, the turret is turned backwards to protect the sensor windows from dirt.

Above left: **Fully loaded Ka-52 with two B-8V-20 rocket launchers suspended on the inner pylons, six Ataka-1 and six Vikhr-1 ATGMs on the middle pylons and two Igla-S anti-aircraft missiles on each outer pylon. Note the tilted barrel of the 30mm gun. (Russian Helicopters)**

Above right: **Six-round UPP-800 block with six tube-launched 9M120-1 Ataka-1 anti-tank missiles with a laser-beam riding guidance. (Piotr Butowski)**

Above left: 9M127-1 Vikhr-1 anti-tank guided missiles with a declared maximum range of 6.2 miles (10km) and 80mm S-8 unguided rockets in front of them. (Piotr Butowski)

Above right: Technicians load the tube with the Vikhr-1 anti-tank missile under the Ka-52 wing of the Korenovsk regiment. (Russian Ministry of Defence)

Armament

The Ka-52 carries weapons and stores on six underwing pylons; four of them have a load capacity of 500kg each, while the two pylons close to the wing tips are used only for Igla anti-aircraft missiles. A cannon is built into the starboard side of the fuselage.

The helicopter has a 9K113U Shturm-VU missile system with 9M120-1 Ataka-1 and 9M127-1 (9A4172K) Vikhr-1 ATGMs, both with laser-beam riding guidance provided by the GOES-451 turret.

The missiles are fired from tube launchers carried in six-round UPP-800 blocks. The UPP-800 has an adjustable position: its nose can be tilted downwards by 10°, thanks to which the missile does not leave the field of view of the targeting sensor after the launch.

The 9M120 Ataka (NATO reporting name: *AT-9 Spiral-2*) anti-tank missile has been developed by the KBM (Konstruktorskoye Byuro Mashinostroyeniya, machine-building design bureau) in Kolomna near Moscow. It officially entered the inventory on 31 May 1996, and series production began a year earlier. The initial version of the missile has a radio command guidance and is not compatible with the Ka-52, which does not have the appropriate equipment (it is used by other Russian combat helicopters, such as the Mi-28N and Mi-24/Mi-35). The 9M120-1 Ataka-1 version with a laser-beam riding guidance, intended for the Ka-52, entered production in 2010 and was officially commissioned on 30 June 2014. Apart from the standard 9M120-1 version with a shaped-charge warhead, there is also the 9M120-1F versions with a high-explosive anti-tank (HEAT) warhead and the 9M120-1F-1 with a high-explosive/fragmentation warhead.

The Vikhr (whirlwind; NATO reporting name: AT-16 *Scallion*) ATGM, developed by the KBP (Konstruktorskoye Byuro Priborostroyeniya, instrument-building design bureau) in Tula, was declared an armament of the Ka-50, and then the Ka-52, for several decades; the work on it began in 1976. Despite the formal commissioning of the first 9A4172 Vikhr version as early as 1990, series production of the missile at the Izhmash plant in Izhevsk began only with the 9M127-1 (or 9A-4172K) Vikhr-1 version in 2013, and on a larger scale from 2018. The only helicopter armed with Vikhr-1 missiles is the Ka-52, although there are plans to equip other helicopters with it; the missile is also being considered as an armament for Inokhodets (Orion) unmanned aircraft.

The Vikhr is launched from a tube; it has an exceptionally elongated body and flies to the target at a high supersonic speed. The guidance is combined: in the first phase of flight the missile is guided by radio commands. It then rides a laser beam sent by the GOES-451 EO turret of the Ka-52 helicopter; integration of Vikhr with the Ka-52 requires increasing the range and accuracy of target recognition by the GOES-451. The pilot of the helicopter can activate the missile's radar proximity fuse before launch and the Vikhr can then combat aerial targets flying at speeds of up to 497mph (800km/h).

Anti-tank Guided Missiles Specifications

Missile	9M120-1 Ataka-1	9M127-1 Vikhr-1
Guidance	laser-beam riding	laser-beam riding
Maximum range	3.7 miles (6km)	6.2 miles (10km)
Maximum range at night	2.8 miles (4.5km)	3.1 miles (5km)
Minimum firing distance	3,281ft (1,000m)	1,640ft (500m)
Armour penetration *)	31.5in (800mm)	31.5in (800mm)
Average speed	783–895mph (350–400m/s)	1,342mph (600m/s) maximum
Calibre	5.1in (130mm)	5.1in (130mm)
Missile length, with tube launcher	72.1in (1,832mm)	113in (2,870mm)
Weight, with tube launcher	107lb (48.5kg)	130lb (59kg)
Warhead weight	16.3lb (7.4kg)	19lb (8.6kg)

*) Rolled Homogeneous Armour equivalent (RHAe)

Against airborne targets, four 9M342 Igla-S (NATO reporting name: SA-24 *Grinch*) or earlier 9M39 Igla-V (NATO reporting name: SA-18 *Grouse*) portable surface-to-air missiles (SAMs) can be carried by the Ka-52, two on each outer pylon. The missiles are carried in 9S846 Strelets twin-round tube launchers; the launcher weighs 143lb (65kg) including the missiles. The KBM Igla man-portable anti-aircraft system, commissioned in 1983, was soon followed by a helicopter-borne version known as the Igla-V, introduced to all Russian combat helicopters. The present 9M342 Igla-S (Super) missile has a two-colour infrared seeker and a rod-type warhead; the range increased from 3.2 miles (5.2km) to 3.7 miles (6km).

KBM 9M342 Igla-S Anti-aircraft Missile Specifications

Guidance	infrared
Maximum range	3.7 miles (6km)
Minimum firing distance	1,640ft (500m)
Average speed	1,342mph (600m/s)
Calibre	2.8in (72mm)
Missile length	5ft 6.5in (1,690mm)
Missile weight	24.9lb (11.3kg)
Warhead weight	5.5lb (2.5kg)

B-8V20A Rocket Launcher Specifications

Weight, empty	220lb (100kg)
Weight, loaded	754lb (342kg)
Length	5ft 10in (1.79m)
Diameter	20.5in (521mm)
Rocket calibre	3.1in (80mm)
Rounds	20

Ka-52 in detail

The firing sequence of the Vikhr-1 missile launched by a Ka-52 of the Korenovsk regiment. (Kalashnikov)

Unguided options of the Ka-52 weapons include up to four B-8V20A launchers, each with 20 x 80mm S-8 rockets. The S-8 is the most widespread Russian unguided air-launched rocket, the first version of which entered inventory in 1973. The basic version, as well as the S-8A (1976) and S-8M with uprated engine (1977) have fragmentation warheads. The S-8KO and S-8KOM have shaped-charge fragmentation warheads. The S-8B and S-8BM versions have sub-calibre (2.7in (68mm)) warheads for concrete constructions penetration. The S-8D (S-8DM) version has a thermobaric warhead. The S-8S has a warhead containing 2,000 arrow-shaped striking elements (flechettes) for fighting manpower. The S-8P is a jamming rocket for the carrier's self-protection with metallised fibreglass chaff dipoles. The S-8-O (S-8-OM) is an illuminating rocket. The S-8 rockets are launched from 20-round B8 launchers in B8M and B8M1 (for fixed-wing aircraft) and B8V20 (for helicopters) versions.

S-8 Rockets Specifications

Version	S-8KOM	S-8BM	S-8DM	S-8-OM
Warhead type	Anti-armour fragmentation	Penetrating	Fuel-air	Illuminating
Weight	24.9lb (11.3kg)	33.5lb (15.2kg)	25.6lb (11.6kg)	26.7lb (12.1kg)
Warhead weight	7.9lb (3.6kg)	16.3lb (7.41kg)	8lb (3.63kg)	9.5lb (4.3kg)
Length	5ft 2in (1.57m)	5ft 1in (1.54m)	5ft 7in (1.70m)	5ft 4in (1.632m)
Calibre	3.1in (80mm)	3.1in (80mm)	3.1in (80mm)	3.1in (80mm)
Range	2.5 miles (4.0km)	1.4 miles (2.2km)	1.9 miles (3.0km)	2.8 miles (4.5km)
Capability	15.7in (400mm) of RHA *)	31.5in (800mm) of concrete		

*) RHA – rolled homogeneous armour

Above left: Two 9M342 Igla-S portable anti-aircraft missiles in a Strelets twin-round tube launcher suspended on the outer pylon of a Ka-52. (Piotr Butowski)

Above right: A Ka-52 fires 80mm rockets. '94' was one of the first Ka-52s delivered in 2010 to the centre in Torzhok. (Russian Helicopters)

Left: A fixed P-800 pod with a Shipunov 2A42 30mm cannon is mounted on the starboard side of the Ka-52's fuselage. (Piotr Butowski)

A fixed P-800 gun pod with Shipunov 2A42 30mm cannon with up to 460 rounds is mounted on the starboard side of the Ka-52's fuselage. The cannon is designed to engage lightly armoured and unarmoured ground and air targets. It was developed by the KBP in Tula and originally intended for the BMP-2 armoured personnel carrier. The hydraulic drive moves the barrel -37/+3.5° in elevation; aiming in azimuth is affected by positioning the helicopter itself. The barrel can be moved by +9/-2.5° in azimuth for cannon stabilisation only. The gun carriage is attached to the fuselage frame near the centre of gravity, where the vibration amplitude is lower, which enables better fire accuracy than in helicopters with a gun under the forward part of the fuselage, especially when firing at an angle to the direction of flight.

The 2A42 has the adjustable rate of fire from 200 to 550 rounds per minute and gives the projectile an initial speed of 960m/s; the cannon weight is 254lb (115kg) and the projectile weight is 380–400g. The advantage of the cannon is the possibility of selective ammunition feed. The ammunition supply is placed in two crates on the sides, and the pilot can decide from which crate the ammunition is to be delivered to the cannon, allowing for the choice between armour-piercing and high-explosive ammunition before opening fire.

In real missions, the weapon set is limited by the permissible take-off weight of the helicopter. The basic (nominal) weapon configuration consists of six Ataka-1 ATGMs and 250 cannon rounds, or six Vikhr-1 missiles and 230 cannon rounds. Another configuration is six Vikhr-1 anti-tank missiles, four Igla-S anti-aircraft missiles and 230 rounds. The variant of loading close to the maximum are two UPP-800 blocks with six Ataka-1 missiles each, suspended on the inner pylons, two B-8V-20 rocket launchers on the middle pylons, and 280 cannon rounds.

For ferry flight, at the maximum take-off weight, four drop tanks are suspended.

Self-protection features

The Russians pay a lot of attention to the armour of combat helicopters. Sergey Mikheyev said during the Ka-52 presentation at Le Bourget 2013: 'We were always surprised why the US military did not require cockpit armour when ordering the Apache.' In Russian doctrine, armour for the crew cockpit, engine compartment and main gearbox is 'categorically necessary', he said. The Ka-52 cockpit has armoured windshields that can withstand a 12.7mm bullet. The armour on the cockpit sides can withstand a 20mm bullet; the backrests of the crew seats are also armoured. Small side armour plates were placed on the outside of the cockpit glazing near the crew heads.

The internal space of the helicopter is organised in such a way as to cover the most important aggregates with less important ones. The rotor blades are made of fibre glass, and they retain their strength after being punctured by a 20mm projectile. The rotor control rods that run outwards are resistant to a double hit by a machine-gun bullet. The fuel tanks are filled with polyurethane foam, preventing explosion in the event of a bullet hit.

The Ka-52 has the L370 Vitebsk self-defence suite to protect against attacks by anti-aircraft missiles and fighter aircraft. The system has been developed by NII Ekran institute in Samara. Work on it started back in the times of the USSR – if you know this, the name Vitebsk, a city in Belarus, and not in Russia, is less surprising. Later, however, the Vitebsk programme stopped for many years, as did the development of new Russian combat helicopters. It was only in 2002 that tests of individual components of the Vitebsk on the Ka-50 helicopter began.

In 2005–06, the Russian MoD ordered the Mi-28N and Ka-52 helicopters, and with them, the self-defence systems. However, while the serial production of the helicopters started quickly, Mi-28N in 2008 and Ka-52 in 2010, the Vitebsk system required several more years of testing. The main problem was the high number of false alarms. Starting from 2011, Vitebsk was produced in small batches

Above left: Armour protection of the Ka-52 pilots' cockpit. (Russian Helicopters)

Above right: A steel plate attached to the outside of the cockpit canopy and covering the pilots from the side. (Piotr Butowski)

Self-defence devices of Ka-52 helicopter: 1 – L370-2 ultraviolet missile launch and approach sensors, 2 – L140 laser warning sensors, 3 – N035 warning radar antennas, 4 – L370-5 directional infrared countermeasures, 5 – UV-26M chaff and flare dispensers. (Piotr Butowski)

intended mainly for special helicopters, for example for the Mi-8AMTSh-1 being a VIP variant for the Russian MoD. The Russians launched the large-scale production of the Vitebsk system in June 2015, after which it was widely deployed in the air force.

The L370 Vitebsk is a modular system designed for all Russian combat and transport helicopters in various configurations. The version intended for the Ka-52 is designated L370P2; the export designation is L370PE2 President-S. The L370P2 suite works completely automatically, controlled by an L370-01 module processing the information received from the warning sensors, forming the commands for jamming devices and giving information to the crew; the information for pilots is displayed on the MFPI-6V panel, which is also the system control panel. The most urgent warning data (type and direction of the threat) is given by voice signals.

The Vitebsk system consists of two sections – the information section with warning sensors and the executive section with countermeasure equipment. The information part of the system consists of four L370-2 ultraviolet missile launch and approach sensors developed and manufactured by NPO GIPO in Kazan. They are positioned on the sides of the Ka-52's front fuselage and the tail boom, covering a total of 360° around the helicopter and 'seeing' the launched missile from a distance of up to 9.3 miles (15km).

An additional warning device installed on the Ka-52 helicopters of the Russian Air Force, cooperating with Vitebsk but not integrated into the system, is the L140 Otklik (response) laser-warning receiver. Four Otklik sensors are installed around the fuselage: two windows on the fuselage sides near the main landing gear, one in the front of the right wingtip fairing, and one at the rear, under the rudder. The sensors cover 360° in azimuth and 90° in elevation. The Otklik sensor works in the 0.4–1.1 micrometre wave range and determines the position of the threat with an accuracy of 10°.

Above left: An ultraviolet L370-2 missile launch and approach sensor attached to the side of the front of the fuselage. (Piotr Butowski)

Above right: In the Ka-52E export helicopter, the newer L418-2E variant is installed instead of the L370-2. (Piotr Butowski)

Above left: In front of the fairing at the end of the right wing there is a window for the L140 Otklik laser-warning sensor. Next there are two UV-26M cartridges, each with 32 26mm decoys, and a position light. This egg-shaped fairing on the stick protruding to the side is the SIVPV-52 air data system. (Piotr Butowski)

Above right: Two more L140 sensors, built into the fuselage, look sideways. (Piotr Butowski)

Above left: The fourth L140 laser-warning sensor, looking back, is installed under the rudder. (Piotr Butowski)

Above right: The flat surface in front of the fairing at the end of the left wing is the antenna for the N035 self-defence radar. Next are the decoy launchers, position light, air data system and a four-signal flare launcher. (Piotr Butowski)

Above left: Two more flat side-looking N035 radar antennas are built into the tailplane endplates. (Piotr Butowski)

Above right: The fourth N035 warning radar antenna, looking backwards, is placed under the fairing under the tail boom. The two circles in front are the antennas of the DISS-32-28 Doppler navigation radar. (Piotr Butowski)

Since 2013, the Zagorsk Optical-Mechanical Plant (ZOMZ) in Sergiyev-Posad near Moscow has been making a modernised L140M laser-warning receiver with a higher sensor sensitivity, the frequency range extended to near and middle infrared, better accuracy and a smaller weight and size of the system. However, the Russian military does not finance work on the L140M, which means it is most likely not interested in the laser sensor. Also, on the Ka-52E export helicopters, L140 Otklik sensors are not installed.

Another information system cooperating with the L370 Vitebsk is the N035 radar, which is part of the FH01 radar system as discussed earlier.

The executive part of the Vitebsk suite on the Ka-52 of the Russian Air Force consists of two L370-5 directional infrared countermeasures (DIRCM) installed on the sides of the lower part of the fuselage, just ahead of the main landing gear. The L370-5 module is a rotating ball containing the SP2-1500 lamp

Above left: Sphere of L370-5 directional infrared countermeasures (DIRCM) in the cruise position on the underside of the Ka-52 fuselage. The jamming lamp window is facing up to avoid damage and contamination. (Piotr Butowski)

Above middle: This is how the same L370-5 jammer looks in the working position. (Piotr Butowski)

Above right: New L418-5E DIRCM sensor for upgraded helicopters. (Piotr Butowski)

that generates modulated infrared and ultraviolet radiation within a 7° angle, jamming the infrared seekers of air-to-air and SAMs. Together, both modules cover a sector of 360° x 90°. The L370-5 DIRCM was developed by SKB Zenith from Zelenograd near Moscow and is series-produced by the Stella-K plant in Kazan (initially by the Stella plant in Saratov).

The Vitebsk system also controls launch of the 26mm (1in) PPI-26 flares or PPR-26 radar decoys from the UV-26M chaff-and-flare dispenser system (CFDS) with two 32-round dispensers in each wingtip fairing. The UV-26M dispenser has been developed by GosMKB Vympel design bureau in Moscow.

Not all intentions regarding the L370 Vitebsk self-defence suite have been realised. In 2008–13, the L370V52 Vitebsk-52 system was under development; it was to be completed with the L370-1V52 radar-warning receiver and L370-3V52 electronic jammer, both by Tsentralnoye Konstruktorskoye Byuro Avtomatiki (TsKBA) from Omsk. These plans were later abandoned.

Kamov Ka-52 Hokum-B Helicopter specifications

Engines: 2 x Klimov VK-2500 rated at 1,790kW (2,400shp) each

Principal dimensions:
Rotor diameter: 47ft 4in (14.43m) upper, 47ft 5in (14.46m) lower
Maximum length, rotors turning: 52ft (15.862m)
Fuselage length: 45ft 6in (13.87m)
Width, no rotors: 25ft 8in (7.835m)
Maximum height: 16ft 5in (5.01m)

Undercarriage:
Type: Retractable with twin nose wheels and single main wheels.
Wheel base: 15ft 1in (4.611m)
Wheel track: 8ft 9in (2.67m)

Weights:
Empty: 17,857lb (8,100kg)
Normal take-off: 22,928lb (10,400kg)
Maximum take-off, for combat: 23,819lb (10,800kg)
Maximum take-off, for ferry flight: 26,896lb (12,200kg)

Performance:
Maximum level flight speed: 162kts (300km/h)
Cruise speed: 140kts (260km/h)
Service ceiling: 18,045ft (5,500m)
OGE hover ceiling, normal weight: 13,123ft (4,000m)
OGE hover ceiling, maximum combat weight: 11,483ft (3,500m)
Maximum climb rate at sea level, 70kts (130km/h)
Speed, normal weight: 3,150ft/min (16m/s)
G limit: 3.0
Practical range at normal weight, 5% reserve: 286 miles (460km)
Ferry range, 5% reserve: 684 miles (1,110km)

Chapter 7

Ka-52E for export

There was no export order for the Ka-50 and Ka-52 for a long time, although the Russians tried very hard. The helicopter was presented abroad many times, starting with Farnborough in September 1992 and Le Bourget in June 1993. Then it participated in the ATAK competitions in Turkey (Ka-50-2 Erdoğan) and AH-X in South Korea (Ka-52K). Talks continued with Algeria, Jordan, Libya, Syria, Yemen, Venezuela and other countries – all without success.

The first foreign presentation of the Ka-52 in its modern configuration took place at Le Bourget in June 2013; '063' participated in the exhibition. Interestingly, the Ka-52 was introduced at Le Bourget with Eurocopter Tiger attack helicopter sensors and armament, including the Safran STRIX optoelectronic turret, PARS 3 LR anti-tank missiles and Mistral ATAM anti-aircraft missiles. In addition, there was the Marte Mk 2/S anti-ship missile.

Soon after, the first – and so far, only – foreign customer was found: Egypt ordered 46 Ka-52E helicopters with delivery in 2017–19. The first helicopter was handed over to the pilots of the newly created 111th Air Wing of the Egyptian Air Force in summer 2017. The Russians are also counting on selling Ka-52K ship-borne helicopters for the Egyptian Navy's Mistral-class ships.

Russia was holding advanced negotiations regarding the sale of 12 Ka-52Es to Algeria; in September 2015, a presentation by a Ka-52 was conducted at Aïn Oussara Air Base in Algeria, including a display of its combat capabilities. So far, however, the contract has not been signed.

In November 2021, test Ka-52E c/n 35383000001, without a side number, participated in the Dubai Airshow, which proves the Middle East and North Africa region remains very important for the Russian helicopter company.

Above left: A performance of Ka-52 '063' at Le Bourget in June 2013. (Piotr Butowski)

Above right: Ka-52 '063' introduced at Le Bourget 2013 with Eurocopter Tiger attack-helicopter sensors and armament, including the Safran STRIX optoelectronic turret, PARS 3 LR anti-tank missiles and Mistral ATAM anti-aircraft missiles. (Piotr Butowski)

Above left: MBDA Marte Mk 2/S anti-ship missile shown together with Ka-52 in Le Bourget, June 2013. (Piotr Butowski)

Above right: An export Ka-52E variant, c/n 35383000001, was shown in Dubai in November 2021, along with options for further equipment upgrades. (Piotr Butowski)

The Ka-52E export helicopter ('izdeliye 800.30') differs from the Russian version with new or modified equipment. Instead of the GOES-451 EO aiming turret, a new OES-52 device, developed by the NPP SPK company of Moscow, was installed. According to Russian media, the prototype for the OES-52 was the French Safran STRIX sight from the Tiger combat helicopter. The OES-52 performs similar functions to the GOES-451 and also houses five sensors: a thermal imaging camera, a TV camera, a laser rangefinder and a target marker, an LSN-02 laser-beam riding ATGM guidance system and a laser spot-detection sensor.

However, it is smaller and lighter than the GOES-451; it weighs 390lb (177kg) compared to 485lb (220kg) previously. The OES-52 turret began trials on a Ka-52 in January 2015 and entered series production in 2017 for the export Ka-52E. It is not known if it will ever be used in helicopters for the Russian military; it is not planned for now, and the Russians have decided to upgrade the GOES-451 turret.

Above left: The Ka-52E has a new smaller OES-52 electro-optical targeting payload in place of GOES-451. Above, you can see the L418-5E MAWS in place of the L370-2. (Piotr Butowski)

Above right: The inscriptions on the export Ka-52E cockpit are in English. Otherwise, it does not differ from the standard cockpit of Russian helicopters. (Piotr Butowski)

Above left: Russian Helicopters Corporation CEO Andrey Boginsky shows the Ka-52E to the Chief of the Mauritanian Air Force, Major General Mohamed Lehretani, during the Dubai Airshow 2021. (Piotr Butowski)

Above right: Laser-guided Kh-38MLE air-to-surface missile laid next to the Ka-52E in Dubai in November 2021. (Piotr Butowski)

Above left: The white button of the L-150-28M.52E Pastel radar warning receiver at the Ka-52E's wingtip fairing. The L140 Otklik laser warning receiver has been removed, and its window is blinded. (Piotr Butowski)

Above right: Two more Pastel's rear-facing sensors are mounted under the Ka-52E rudder. (Piotr Butowski)

Left: The first, as yet unpainted, Ka-52E in trials in Arsenyev in May 2017. (Progress)

The Ka-52E export helicopter retained the standard radar; the export FH01E Arbalet-52 differs from the Russian version only in operating frequencies.

The helicopter's self-defence system (export designation L370PE2 President-S) was significantly modified. The L370-2 missile approach-warning sensors were replaced with more modern L418-2E sensors of the L418 Monoblok system.

The L140 Otklik laser-warning sensors were abandoned and, instead, the L-150-28M.52E Pastel radar-warning receiver produced by TsKBA from Omsk was installed. Two Pastel sensors are placed at

Above: Egyptian Ka-52Es on board the Mistral-class ENS *Gamal Abdel Nasser*, together with AH-64D Apaches. (Egyptian MoD)

Right: Egypt ordered 46 Ka-52E helicopters; deliveries started in summer 2017. (Egyptian MoD)

the wingtip fairings, looking forward, left and right, and two more sensors looking sideways to the rear are placed under the rudder, where the Otklik window is located in Russian helicopters.

The Pastel radar-warning receiver operates in the G–J range (wavelength 1.5–7.5cm, frequency 4–20 GHz); the observation zone is 360° x 60°. The sensors detect ground or airborne radar from a distance 20 per cent greater than the distance the helicopter is detected by this radar, the manufacturer claims.

Why does the Ka-52E export helicopter have newer equipment than the Russian version of the Ka-52?

The Ka-52E export helicopter received the OES-52 optoelectronic turret, which, in the words of the Kamov head, Sergey Mikheyev, was '1.6 to 2 times better' than the GOES-541 installed on the Russian Air Force's helicopters. The new self-defence devices, the L418-2E and L418-5E, had been used on the Egyptian Ka-52E helicopters earlier than the Russian Ka-52 helicopters. Similarly, the Mi-28NE helicopters delivered to Iraq received their radars in February 2015, while deliveries of the Mi-28UB helicopters with radars to the Russian Air Force only started at the end of 2017.

The fact that new equipment is often exported faster than it is delivered to the Russian Air Force is due to more stringent formal requirements for equipment for the Russian MoD. Sometimes, it is also because of additional functions implemented in the equipment for the Russian military: the implementation of a new device in Russia requires a multi-stage trial, which is not necessarily needed by a foreign customer.

Chapter 8
Ka-52K Katran ship-based helicopter

In 2010, for the first and only time in modern history, Russia bought large combat equipment from a NATO country. On 24 December 2010, Russian President Dmitry Medvedev sent a letter to French President Nicolas Sarkozy, informing him that Russia had selected a French bid for the supply of helicopter-carrying ships. The tender had only been announced on 5 October 2010 and was, in fact, purely a formality. Talks with France had already been going on for over a year then, and the discrepancies concerned, as usual, the price and the conditions for technology transfer to Russia. The Russians wanted to buy one ship and build the other three under a French licence; the French insisted on a 2 and 2 split. A large part of the ship's equipment and armament, including the ship-borne helicopters, were to be of Russian production. The tender was a form of pressure in this situation, especially since it was not known if anyone would participate in it apart from the French, although the Russian Defence Minister Anatoly Serdyukov said that companies from Germany, Spain and South Korea had been invited to bid.

Even earlier, in November 2009, a French Mistral ship visited St Petersburg, where it was shown to the commanders of the Russian Navy; Kamov Ka-27, Ka-29 and Ka-52 helicopters landed aboard the ship.

The formal contract was signed in June 2011; Russia ordered two Mistral helicopter carriers be delivered from France, with an option for a further two to be built in Russia. The Russian media, citing their sources participating in the negotiations, announced the price for the first ship was to be €720 million, and the second would be €650 million, including spare parts and crew training.

The ships were to be built by the STX France Chantiers de l'Atlantique shipyard in Saint-Nazaire in cooperation with the Russian Baltiysky Zavod shipyard in St Petersburg, which sent their parts to France for final assembly. The Russians' share in the construction of the first ship was to be 20 per cent, in the second it could reach 40 per cent, if the Russians were able to master the new process. The first ship, *Vladivostok*, was launched on 15 October 2013 and began its sea trials on 5 March 2014. In the Russian Navy, it was to belong to the Pacific Fleet and be based at Vladivostok. The second ship was to be called *Sevastopol*, due to be ready a year later, to serve in the Black Sea Fleet.

This is what *Vladivostok*, one of the four Mistral-class landing helicopter docks for the Russian Navy, was supposed to look like. (DCNS)

Above left: In November 2009, a French Mistral ship visited St Petersburg, where Kamov Ka-27, Ka-29 and Ka-52 helicopters landed aboard. (Alexey Mikheyev, Russian Helicopters)

Above right: In 2015, Ka-52 '062' had been converted into a technology demonstrator for a Ka-52K shipborne version. Note the new short wing. (Russian Helicopters)

The Mistral-class ship is classified in France as a bâtiment de projection et de commandement (force projection and command ship), and in the NATO classification it is a 'landing helicopter dock'. It has a full load displacement 21,300 tonnes, a speed of 18.8 knots and a range up to 10,700nm (19,800km). It can carry 450 soldiers of the assault force and up to 40 tanks. An air group of, typically, eight NH90 landing helicopters and eight Tiger combat helicopters may be stationed on a Mistral; a maximum of 30 helicopters can be deployed on the ship. The Mistral-class ship has six landing spots for helicopters on deck, as well as two elevators of a 13-tonne load capacity, linking the deck with the hangar under it. The ship has been slightly modified to meet the specific Russian requirements. The hull is strengthened to withstand thin ice. The aircrews' and soldiers' compartments are more modest. Moreover, as the Kamov Ka-29 helicopters are 17ft 9in (5.4m) high (4ft 4in [1.3m] more than the NH90 on parking) the ceiling in the helicopter hangar has been raised.

The Kamov company was the one who was most pleased with the Russian Navy's purchase of a Mistral. 'For us, helicopter manufacturers, the purchase of Mistral is great news,' said Russian Helicopters' deputy CEO, Andrey Shibitov, adding that, thanks to this, 'a new life enters the development of Russian ship-based helicopters'. The most important element of the air group aboard the Russian Mistrals were to be modernised Ka-29M transport and landing helicopters supported by Ka-52K combat helicopters. The deployment of Ka-31 early-warning helicopters was also planned, and a modern search and rescue helicopter was to be built on the basis of the Ka-32A-11BC civilian helicopter. A light Ka-226 helicopter was provided for liaison and auxiliary tasks.

A ship-based combat helicopter, Ka-52K (Korabelnyi, ship-borne; 'izdeliye 800.20'), nicknamed *Katran* (spiny dogfish), was developed especially for Mistral helicopter carriers. Ka-52KUB (Uchebno-Boyevoi, combat training) was its planned trainer version. To save space on board, the Ka-52K has a modified rotor hub, which enables the blades to be folded. Also, the wing of the sea helicopter is shorter than in the land-based version. Additionally, when parked, it is folded (tilted back with a hinge on the rear edge near the weapons pylon); in the first design, shown on an earlier model, the outer wing panel was raised up. The undercarriage is reinforced to withstand being landed on a moving deck. Inflatable floats, in case of emergency ditching, can be fitted on the fuselage sides.

Especially for ship-borne operations, the new navigational equipment and instrumental landing system is mounted. Ultimately, the plan is to install a radar operating in the X-band (3cm wavelength), which will achieve a much greater range than the current Ka-band (8mm wavelength) radar.

Above left: Head of the Kamov company, Sergey Mikheyev, with the first Ka-52K during the Army exhibition in Kubinka in June 2015. (Piotr Butowski)

Above right: The first of four pre-production Ka-52K helicopters, c/n 35382001001, made its maiden flight on 7 March 2015, and a few months later, it was demonstrated at the Army 2015 in Kubinka near Moscow. (Piotr Butowski)

The resolution of the X-band radar is smaller than that of the Ka-band, but it is sufficient against sea and air targets. For now, the Ka-52K helicopter has the same armament as the land Ka-52, i.e., Vikhr and Ataka anti-tank missiles, unguided rockets and a standard 30mm cannon. Due to the shorter wing, the Ka-52K only has four pylons instead of the Ka-52's six. Eventually, the Ka-52K's weapons suite is to be supplemented by Kh-35UV anti-shipping missiles and Kh-38M air-to-ground missiles. The Ka-52K has been presented with these missiles several times, but they have not yet been integrated into the helicopter's mission suite.

In 2010–15, the Ka-52K programme was a priority for Kamov. At the very end of August 2011, as part of the work on the seaborne version of the helicopter, the prototype Ka-52 '061' made several take-offs and landings on board the *Vice-Admiral Kulakov* (Udaloy-class) destroyer in Severomorsk. In 2012, an initial batch of four Ka-52K helicopters was ordered from the Arsenyev plant, while the Kamov workshop in Lyubertsy converted the second Ka-52 prototype, '062', into a technology demonstrator for the Ka-52K, with a new folding wing and folding rotors; the modified helicopter took off for the first time in January 2015.

Above left: In August 2015, the same first Ka-52K was presented with the new OES-52 optoelectronic turret. Note the folded wing and rotor blades, as well as a block of six Ataka-1 anti-tank missiles. (Piotr Butowski)

Above right: The Ka-52K's short wing, with a pack of six Vikhr-1 missiles and a launcher of 20 80-mm rockets underneath. (Piotr Butowski)

Above left: In the fairing at the end of the right wing of the Ka-52K, in addition to the usual self-defence decoy launchers, there is an orange 'black box' – the emergency flight data recorder. (Piotr Butowski)

Above right: Ka-52K '103' helicopter cockpit. (Piotr Butowski)

Right: Ka-52K '103', which took part in the expedition to Syria on board the *Admiral Kuznetsov* aircraft carrier between October 2016 and February 2017. (Piotr Butowski)

The first of four pre-production Ka-52K helicopters, c/n 35382001001, still without the side number, made its maiden flights on 7 March 2015 in Arsenyev, and a few months later, the aircraft was demonstrated at the Army 2015 forum in Kubinka near Moscow. All four helicopters were ready shortly thereafter.

On 8 April 2014, the Russian MoD placed an order for 32 Ka-52K helicopters (not including the four test ones), the first 12 of which were to be delivered in 2015. In 2013, Ka-52K flight simulators were also ordered for the 859th Combat Training and Flight Crew Conversion Centre of Naval Aviation in Yeysk and for the 289th Independent Composite Aviation Regiment of the Pacific Fleet in Nikolayevka near Vladivostok.

Russian aggression against Ukraine in the spring of 2014 changed everything. On 3 September 2014, French President François Hollande announced France was suspending the delivery of the two Mistral-class ships to Russia, and on 25 November, the contract was cancelled. In October 2015, both ships were acquired by Egypt. The first entered service with the Egyptian Navy in June 2016 as ENS *Gamal Abdel Nasser*, and the second, three months later, as ENS *Anwar El Sadat*.

Thus, further work on the Ka-52K lost its raison d'être, at least for a while. The management of the Russian Helicopters corporation declared that, despite the lack of Mistrals, the Ka-52K programme would continue unchanged as the helicopter could be used 'on other Russian ships as well as by Navy coastal aviation'. It was putting

In memory of the Syrian expedition, the emblem 'Za dalniy pokhod' ('For the far crusade') was painted on Ka-52K '103' cockpit's side. (Piotr Butowski)

Ka-52K '104', spotted in August 2019, had an OES-52 targeting turret, L418-2 and L150 Pastel warning sensors as well as L418-5 jammers. Under the wing hangs a container with recording equipment, which has been typical for Kamov helicopters undergoing tests for several decades. (Piotr Butowski)

Ka-52K '104' performs with slightly different equipment during a navy parade in St Petersburg in July 2021; in particular, it has a GOES-451 sighting turret. The helicopter is armed with six Ataka-1 missiles, six Vikhr-1 missiles and two rocket launchers. (Piotr Butowski)

on a happy face, though, as the Russian Navy does not have any ships on which such a helicopter would be needed. Russian warships – missile cruisers, destroyers and frigates – each carry one or two anti-submarine or search and rescue helicopters and do not need the Ka-52K attack helicopters. In May 2015, Andrei Shibitov, the deputy CEO of Russian Helicopters, admitted that 'for today, there is no order' for Ka-52K helicopters; it is possible the earlier 32-helicopter contract has been cancelled.

The four Ka-52K helicopters of the first series are currently being used by Kamov to test new systems for the Ka-52M and Ka-52E programmes. For example, at the MAKS 2015 exhibition, the first helicopter had an OES-52 optoelectronic targeting turret; earlier, at Army 2015, and later at MAKS 2017 exhibitions, the same helicopter had the standard GOES-451 turret. Ka-52K '103' with a new Zaslon RZ-001E Rezets active electronically scanned array (AESA) radar was shown at MAKS 2019 and 2021 exhibitions. Ka-52K '104' performed at MAKS 2019 with the OES-52 turret, the Pastel radar-warning receiver and the L418-5 infrared countermeasures sensors; two years later, in July 2021 at the maritime aviation parade in St Petersburg, the same helicopter had a GOES-451 turret.

From November 2016 to January 2017, Russians deployed their only aircraft carrier, *Admiral Kuznetsov*, to the Syrian coast, primarily to gain experience. The military significance of this mission was minimal, as all the tasks were easier and more effective to perform from the Russian Hmeimim land base in Syria. Even the MiG-29K and Su-33 ship-borne fighters, when in Syria, performed 90 per cent of their missions from this land base. On board the ship, among the many types of aircraft, there was one Ka-52K. The helicopter's number was painted over for the duration of the mission, but it was later restored. Moreover, in memory of this expedition, the emblem 'Za dalnyi pokhod' ('for the far crusade') was painted below the cockpit; hence we know that it was '103'. There is no information about the activities of the Ka-52K during the deployment to Syria.

Above left: This is what the future universal landing ship *Ivan Rogov*, Project 23900, is planned to look like. The ship is intended to carry up to 18 Ka-52K and Ka-29 helicopters. (AK Bars)

Above right: The Ka-52K's armament set is to be expanded with the Kh-35UV anti-shipping missile. Here, the missile is presented with '062', which is a technology demonstrator for the K version. (Kamov)

Good news for Ka-52K came on 20 July 2020. At the Zaliv shipyard in Kerch in the Crimea, in the presence of Vladimir Putin, the keel was laid for two new universal landing ships – Project 23900. The helicopter carriers are to have a standard load displacement of 23,000 tonnes and take on board up to 18 helicopters, including the Ka-29M transport and the Ka-52K combat aircraft. The first ship, *Ivan Rogov*, is to enter service in 2028 in the Pacific Fleet, and the second, *Mitrofan Moskalenko*, a year later in the Black Sea Fleet. The commander-in-chief of the Navy, Admiral Nikolai Yevmenov, told the Russian press that these ships outperform the French Mistrals with their 'seaworthiness as well as operational and design characteristics'.

The Russians are actively promoting the Ka-52K in Egypt after the country bought the two Mistral-class ships originally manufactured for Russia, as well as 46 'regular' land Ka-52Es. The purchase of the Ka-52K by China is also possible, as evidenced by a Chinese television report from the Arsenyev plant in September 2021. In 2018, China Shipbuilding Industry Corporation presented a model of its own 'landing ship personnel and dock' with Ka-52K helicopters on board.

Kamov Ka-52K Helicopter Specifications

Engines: 2 x Klimov VK-2500 or VK-2500P rated at 1,790kW (2,400shp) each

Principal dimensions:
Rotor diameter: 47ft 4in (14.43m) upper, 47ft 5ins (14.46m) lower
Maximum length, rotors turning: 52ft (15.862m)
Fuselage length: 45ft 7in (13.90m)
Wingspan: 20ft 7in (6.303m)
Maximum height: 16ft 9in (5.10m)

Undercarriage:
Type: Retractable with twin nose wheels and single main wheels.
Wheel base: 15ft 1in (4.611m)
Wheel track: 8ft 9in (2.67m)

Weights:
Maximum take-off weight: 26,896lb (12,200kg)
Maximum weapons and stores: 4,409lb (2,000kg)

Performance:
Maximum level flight speed: 157kts (290km/h)
Cruise speed: 135kts (250km/h)
Service ceiling: 17,060ft (5,200m)
OGE hover ceiling: 11,810ft (3,600m)
Maximum climb rate at sea level: 2,756ft/min (14m/s)
G limit: 2.5
Practical range: 280 miles (450km)
Ferry range: 684 miles (1,100km)

KTRV Kh-35UV Anti-shipping Missile

In March 1983, following an assessment of the Falklands War, the Soviet Union began to develop its own equivalent of the Exocet and Harpoon missiles. The Tactical Missiles Corporation (KTRV) Kh-35 ('izdeliye 78'; NATO reporting name: AS-20 *Kayak*) was originally developed as a naval anti-shipping missile for ship and coastal launches; a first test launch was made from a shore-based position on 5 November 1985. The Kh-35V helicopter-borne missile was first fired from a Ka-27 in December 1991, but trials ceased after only three launches due to lack of money.

Kh-35U ('izdeliye 07'; for export: Kh-35UE) is a considerably modernised missile being developed in ship-, shore-, helicopter- and aircraft-launched versions. The airborne missile was launched for the first time on 1 November 2010 by a Su-34 fighter-bomber; the missile is in series production.

The Kh-35UV (Vertolyotnaya, helicopter) is a sub-variant for Ka-52K helicopters and fitted with a Kartukov Iskra 78DTV rocket booster to enable launch at low speed. The cruising propulsion is provided by a Saturn 'izdeliye 64M' turbofan engine. The missile approaches the target at high subsonic speed and very low altitude. Guidance is provided by a Ts-074U inertial navigation system and a satellite navigation receiver in cruise flight, while an active-radar ARGS-35U (U-502U) seeker is used for the terminal phase; the seeker has a publicised lock-on range of 31 miles (50km). The missile has a semi-armour-piercing/fragmentation warhead.

Kh-35UV Missile Specifications

Maximum range	162 miles (260km)
Cruising altitude	33–49ft (10–15m)
Terminal flight altitude	13ft (4m)
Maximum speed	Mach 0.85
Launch weight	1,433lb (650kg)
Warhead weight	320lb (145kg)
Length, with rocket booster	14ft 5in (4.40m)
Body diameter	16.5in (420mm)
Wingspan	52.4in (1.33m)

Chapter 9
Modernised Ka-52M

Each aircraft or helicopter in Russia is eventually given the letter 'M' (for 'modernised') in its designation. On 5 April 2019, Kamov received a state contract from the MoD for the Avangard-4 R&D work, to develop a modernised Ka-52M helicopter; at the same time, the modernised Mi-28NM was being built by Mil under the codename 'Avangard-3'. The actual works started much earlier than the contract, because the new equipment and weapons used in the Ka-52M were then almost ready.

In June 2020, the Mil and Kamov National Centre of Helicopter Engineering (this is a new structure created at the end of 2019 from the merger of two Russian helicopter design bureaus; the Kamov company does not formally exist anymore) commissioned the Progress plant in Arsenyev to upgrade two Ka-52 helicopters into test examples of new Ka-52M version. The helicopters, c/n 35382617009 and 35382617010, the ninth and tenth helicopters of the 17th production batch, were selected for conversion.

The first Ka-52M made its first flight on 10 August 2020. In July 2021, the then director general of the Russian Helicopters holding Andrei Boginsky said that the helicopter had completed preliminary tests and was ready to start state evaluations conducted by the MoD. According to the contract, the Ka-52M state trials are to be completed by the end of September 2022, which means that the helicopter will then be ready for serial production.

The Ka-52M was shown to the public for the first time, but only from a distance, during the MAKS airshow in July 2021. The s/n 17-10 participated in the flight display taking off from Tomilino airfield belonging to Russian Helicopters, 15.5 miles (25km) from Zhukovsky. It was shown up close a month later, during the Army 2021 forum in Kubinka.

As part of the state armament programme for 2018–27 (GPV-2027), the Russian Federation MoD intends to purchase 114 modernised Ka-52 helicopters, said Yuri Borisov, the then deputy defence minister, during his visit to Arsenyev in February 2018. This was repeated by his successor, Alexei

One of the two Ka-52M helicopters, c/n 35382617010, used for trials that appeared in the air during MAKS 2021 in Zhukovsky. The new GOES-451M sight is in the working position. (Piotr Butowski)

Krivoruchko, who visited Arsenyev on 28 May 2019; he announced the signing of an appropriate contract for 2020. The future contract will also include an upgrade of the previously produced Ka-52s to Ka-52M standard. The contract for the first batch of 30 Ka-52M helicopters, within the planned 114, was signed on 14 August 2021, during the Army 2021 forum. Fifteen helicopters are to be delivered in 2022 and 15 in 2023. The contract was signed on behalf of the MoD by Deputy Minister Alexei Krivoruchko and, on behalf of the Russian Helicopters holding company, by Andrei Boginsky.

The Ka-52M ('izdeliye 800.50') helicopter received the modernised GOES-451M EO targeting turret with an increased detection and recognition range, the modernised BKS-50M (Bortovoi Kompleks Svyazi, onboard communications complex) communication suite, as well as the modernised SUO-806PM (Sistema Upravleniya Oruzhiyem, weapon control system) stores management system, capable of managing new weapons. The most important novelty in the helicopter's armament is the implementation of the LMUR (Lyogkaya Mnogofunktsionalnaya Upravlayemaya Raketa, lightweight multipurpose guided missile) with a range of up to 9 miles (14.5km). There were also upgrades introduced to the helicopter itself. The Ka-52M rotor blades have a more powerful heating element, which enables the helicopter to operate in a wider temperature range, including in arctic conditions. The helicopter received undercarriage wheels with increased load capacity and wear resistance, as well as external LED lighting. The crew cockpit has improved ergonomics and is also better adapted for piloting in the dark with night-vision goggles.

Kamov Ka-52M Helicopter Specifications

Weights:
Maximum take-off weight: 26,896lb (12,200kg)
Performance:
Maximum level flight speed: 162kts (300km/h) at 22,928lb (10,400kg) weight
Cruise speed: 140kts (260km/h) at 22,928lb (10,400kg) weight
Service ceiling: 18,045ft (5,500m) at 22,928lb (10,400kg) weight
Practical range: 286 miles (460km)

Above left: The Ka-52M at MAKS 2021. Interestingly, the self-defence suite is the standard L370 Vitebsk and not the modernised L418 Monoblok. (Piotr Butowski)

Above right: The Ka-52M at MAKS 2021. This odd-shaped pod at the end of the left wing with a radio-transparent nose houses the AS-BPLA command line for communication between the helicopter and the LMUR missile. (Piotr Butowski)

New aiming sensors

The GOES-451M EO payload, currently being tested on the Ka-52M prototypes, is heavier than the original GOES-451, but this time the range was chosen as a priority, to match the range of the new types of weapons, including the LMUR missile. The manufacturer's declared range of the tank detection is 9.3 miles (15km) via the TV sensor and 7.5 miles (12km) by the thermal imaging sensor; the target recognition ranges are 7.5 miles (12km) and 5 miles (8km), respectively. The GOES-451M has additional sensors (the 'M' version has eight various-sized windows for sensors, compared to six in the GOES-451), but their specific purpose is unknown.

For now, the Ka-52M has retained the current Phazotron-NIIR FH01 radar, but there are several options being considered for the future. The most actively promoted by the Russians is the new V006 Rezets radar from the Zaslon company of St Petersburg (the same company makes the V004 radar for the Su-34 fighter-bomber). Rezets was shown for the first time at the MAKS 2019 exhibition, where Ka-52K '103' with a test copy of this radar was present. Separately from the helicopter, the radar was shown at the Dubai Airshow in November 2021.

The V006 radar, presented to the public with the export designation RZ-001E Rezets (cutter), has a fixed AESA antenna, 900mm x 300mm in size, with 640 transceiver modules. The radar works in the X-band, has the pulse power of 1.8kW and, according to the manufacturer, can detect a group of tanks from 25 miles (40km), a railway bridge from 62 miles (100km) and a destroyer-class warship from 93 miles (150km). In the air-to-air mode, it detects a fighter aircraft (radar cross-section, RCS, $3m^2$) from up to 31 miles (50km) head-on and 12 miles (20km) tail-on. The radar weighs 287lb (130kg) (22lb [10kg] less than the present FH01). The Rezets radar is air cooled and a new air scoop for radar cooling appeared on the helicopter's nose fairing.

Also Phazotron-NIIR, the manufacturer of the current FH01 radar, offers several variants of their radars for the Ka-52M. The modernised FH02 radar would receive, in addition to the current Ka-band (8mm), another X-band (3cm) channel. In the basic option, the manufacturer offers two separate antennas – a mechanical slotted one for the Ka radar and an electronic one for the X radar. This solution enables simultaneous observation of the surface and airspace, implementing advanced target detection and tracking algorithms and achieving greater reliability. The X-band can achieve much longer range, although at the expense of lower resolution. According to the company, a railway bridge can be detected at a distance of 20 miles (32km) in the Ka range or 78 miles (125km) in the X range, and a tank can be detected at 12 miles (20km) or 22 miles (35km), respectively. The angular resolution is 0.8° for the Ka channel, but only 3.2° for the X channel. Additionally, the radar can detect airborne targets (RCS $5m^2$) from a distance of 56 miles (90km).

Russian Helicopters CEO Andrey Boginsky and Defence Vice Minister Alexei Krivoruchko have just signed a contract for the supply of 30 Ka-52M helicopters, 14 August 2021. (Russian Helicopters)

As a low-cost solution, Phazotron-NIIR proposed to use in FH02 the current parabolic mechanical antenna from the FH01, as it is common for both channels. Such a variant is cheaper and easier to make but has limited tactical capabilities, as only one channel can work at a time, either Ka or X.

Another radar, offered by Phazotron-NIIR primarily for the further modernisation of the Ka-52K seaborne helicopter, is the FH03. Work is carried out by Phazotron on its own initiative and for its own money. The X-band radar is supposed to get a fixed front AESA antenna made in the shape of a half-cylinder; therefore, the radar viewing angles will be +/- 85° in azimuth.

Apart from the upgrade of the main nose radar, an upgrade of the L-band N035 radar is also planned; as previously mentioned, the N035 has four small antennas and is used as a missile approach-warning sensor. After modernisation, it would be used for area mapping and detecting ground objects. The Ka-band radar would remain the basic sensor for a target attack at a distance up to 5 miles (8km) and for reconnaissance and target recognition at a distance of 19–25 miles (30–40km), while the L-band radar would be used primarily for reconnaissance at a distance of up to 81–93 miles (130–150km).

The software of the standard FH01 radar is constantly being improved, including the implementation of the synthetic-aperture mode (which improves the resolution of the radar, but it only works in flight at a speed of at least 54kts [100 km/h]), as well as the detection of hovering helicopters.

New missiles

The most important new armament of the modernised Ka-52M is the LMUR, or 9A-7755, or 'izdeliye 305' ('305E' is an export version). The missile was developed at the KBM design bureau in Kolomna, which also made the Ataka ATGM used on Ka-52.

The LMUR missile is closer to a helicopter-launched loitering munition than to a regular ATGM of the Ataka type. The range of the LMUR is up to 9 miles (14.5km), while the current Ataka or Vikhr missiles reach 3.7–5 miles (6–8km). In the simplest way, the LMUR missile is fired against a target that has been locked by the missile seeker before launching. In such a variant, the range of the missile is limited by the capabilities of the seeker.

In the second mode, first used in the Russian anti-tank missiles, the LMUR can be fired at a target that is invisible, far away or hidden. First, the missile flies to the target area guided by an inertial

Above left: The modernised GOES-451M is heavier than the original GOES-451, but this time the range was chosen as a priority to match the range of the new types of weapons. The manufacturer's declared range of the tank detection is 9.3 miles (15km). (Piotr Butowski)

Above right: The new AESA RZ-001 Rezets radar has been tested on Ka-52K '103' since 2019. It can detect a group of tanks from 28 miles (45km) and a hovering helicopter from 12.4 miles (20km). (Piotr Butowski)

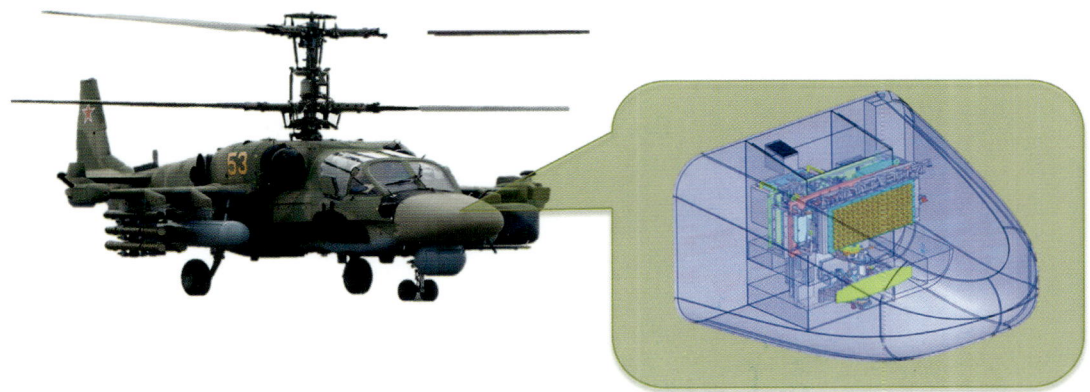

Above: The twin-antenna FH02 radar proposed for Ka-52M by the Phazotron company. (Phazotron)

Right: Another radar option for Ka-52, primarily for the seaborne version, is the FH03 with a half-cylinder-shaped AESA antenna looking +/- 85° in azimuth. (Phazotron)

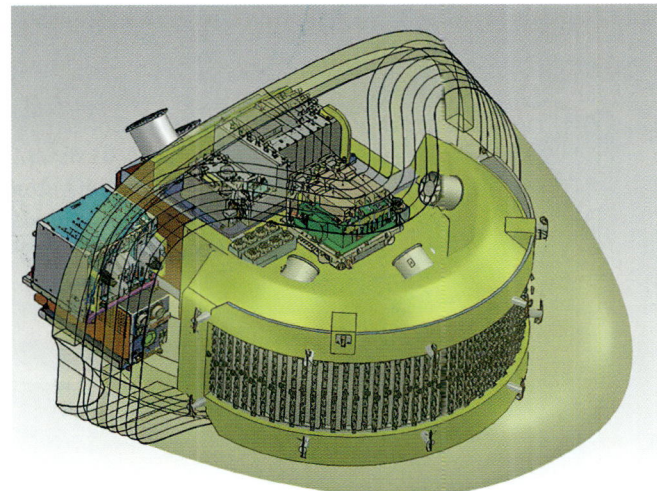

autopilot with correction by the satellite navigation receiver. The operator on board the helicopter sees the image from the missile's optical/thermal imaging 9B-7755 seeker and makes the selection and indication of a specific target to the missile. Communication between the helicopter and the missile is provided by the AS-BPLA (Apparatura Svyazi s Bespilotnym Letatelnym Apparatom, unmanned aerial vehicle communication equipment) command line installed in a container under the wing. In this mode, it is possible to use the maximum range of the missile. It is also possible to indicate targets by other means, such as unmanned aerial vehicles.

The 'izdeliye 305' missile has a canard aerodynamic configuration, with a folding wing. A practice variant of the missile, 'izdeliye 305-UL' (Uchebno-Lyotnaya, flight training), differs from the combat version with a smaller, non-foldable wing.

The LMUR programme was launched on 22 February 2011, by order of the Russian MoD; the R&D work received the codename 'Prefix', and the missile 'izdeliye 79'. However, after a few years, the unfinished programme '79' was taken over by the FSB, which wanted to have a long-range weapon for their helicopters used in special operations in the Caucasus. With the takeover of the project by the FSB, the requirements for the missile, now called 'izdeliye 305', were supplemented with the video

An LMUR 'izdeliye 305' guided missile carried on an APU-305 rails under the Ka-52's wing. The missile seeker is covered with a cap that protects it from dirt and is raised directly before the launch. It is actually a practice version of the missile with a smaller wing. (Piotr Butowski)

A combat version of the LMUR missile with a large folding wing. The missile range reaches up to 9 miles (14.5km). (Michael Jerdev)

The new Kh-MD, or 'izdeliye 85', modular tactical missile (the equivalent of the British-French MBDA Sea Venom/ANL) was displayed next to the Ka-52M at the Army 2021 exhibition. (Michael Jerdev)

The 9M123 Khrizantema (above) ATGM is a further development of the 9M120 Ataka (below) with a more powerful warhead. (Piotr Butowski)

The KBP Kornet anti-tank missile is planned for Ka-52 in an extended-range Kornet-DA version able to reach 6.2 miles (10km). The Kornet uses the same laser-beam riding guidance as the Ka-52's other missiles. (KBP)

transfer from the missile seeker to the helicopter and the target indication by the operator; the FSB wanted to be able to control the situation to the end and possible 'recall' of the missile, if the target turned out to be different than expected.

The LMUR missile has been in series production since around 2016 and was first deployed on the Mi-8MNP-2 helicopters operated by the FSB. Since 2019, the LMUR has been tested on the modernised Mi-28NM combat helicopter, and from 2020 also on the Ka-52M. The Ka-52M helicopter typically takes four missiles on APU-305 rail launchers, and a maximum of eight missiles on APU-L double rails.

KBM LMUR (305E) Missile Specifications

Maximum range	9 miles (14.5km)
Maximum speed	447kts (230m/s)
Flight altitude	328–1,969ft (100–600m)
Weight	231lb (105kg)
Warhead weight	55lb (25kg)
Length	76.6in (1,945mm)
Body diameter	7.9in (200mm)

Another new missile for the Ka-52M is the Kh-MD (Maloi Dalnosti, short range), or 'izdeliye 85', made by the Tactical Missiles Corporation (Korporatsiya Takticheskoye Raketnoye Vooruzheniye, KTRV), the main manufacturer of aviation weapons in Russia. At the Army 2021 exhibition, the Kh-MD missile was displayed next to the Ka-52M.

According to the manufacturer, the Kh-MD, referred to as MMM ASP (Mezhvidovoye Malogabaritnoye Modulnoye Aviatsionnoye Sredstvo Porazheniya, joint small-size modular air-launched weapon) is a universal, modular tactical missile, the equivalent of the British–French MBDA Sea Venom/ANL missile. It is available in three versions: Kh-MD-E1, E2 and E3 (E stands for the export variant), which differ only in the type of the seeker; at the initial stage of flight, inertial navigation combined with satellite navigation is implemented in all versions. The missile shown with the Ka-52M helicopter had a gyrostabilised laser seeker. Another version has a 3mm active-radar seeker by the Mitinopribor company, as well as an additional infrared sensor, which increases the accuracy and effectiveness of aiming.

KTRV Kh-MD-E Missile Specifications

Range when launched from ground	12.4 miles (20km)
Range when launched from helicopter	24.9 miles (40km)
Weight	243lb (110kg)
Warhead weight	66lb (30kg)
Length	94.5in (2,400mm)
Body diameter	7.9in (200mm)

According to reliable information, the Ka-52 is also to be armed with two other types of anti-tank missiles, the Khrizantema (chrysanthemum) and the Kornet. They have long been used by ground troops, but so far have not been presented with the Ka-52.

A product of KBM, the 9M123 Khrizantema (NATO reporting name: AT-15 *Springer*) ATGM is a further development of the Ataka. The Khrizantema-S system, mounted on an infantry combat vehicle chassis, entered service with the Russian ground forces in 2005. The missile reaches the same 3.7 miles (6km) range and 895mph (1,440km/h) average speed of the Ataka but has a more powerful 6in (152mm) over-calibre warhead (the warhead diameter is greater than body diameter) able to penetrate 43–47in (1,100–1,200mm) RHAe (Rolled Homogeneous Armour equivalent). The missile's unique feature is the availability of two guidance channels: semi-automatic laser-beam riding guidance (anti-tank 9M123 and blast 9M123F missile) and automatic radar homing (9M123-2, 9M123F-2). Thanks to this, two missiles can be simultaneously guided towards two separate targets. Using the radar guidance, the missile can be employed at night, in difficult weather conditions (fog, rain or snow) or under a smokescreen. However, using the radar guidance requires the helicopter to be equipped with an additional radar operating in the frequency range 100–150 GHz (wavelength 2–3mm), through which the missile is automatically guided to the target; such a radar is known in the container version for the Mi-28N, but has never been presented on the Ka-52.

The second missile is the 9M133M Kornet (NATO reporting name: AT-14 *Spriggan*) from the KBP design bureau in Tula, the same that made the Vikhr missile for the Ka-52. The Kornet is planned for the Ka-52 in an extended-range version of the Kornet-DA, which flies up to 6.2 miles (10km) at a speed of 320m/s (a standard Kornet reaches 3.4 miles [5.5km]). The weight of the missile, together with the tube launcher, is 73lb (33kg), and the length is 1,210mm. The Kornet missile has semi-automatic laser-beam riding guidance, so it uses the same GOES-451 aiming sensor as the other Ka-52 missiles.

New self-defence system

For the Ka-52M helicopter, as with other new Russian helicopters, the NII Ekran institute in Samara is developing the L418 self-defence system made within an R&D project codenamed Monoblok, which launched at the beginning of 2010. The L418 is a modernisation of the L370 Vitebsk system, the first goal of which was to reduce weight and costs of the devices thanks to the use of a new element base. The modernisation programme received an additional impetus with the Russian operation in Syria, where the Vitebsk system was used on the Ka-52, Mi-8AMTSh and Mi-35M helicopters. In March 2017, Deputy Defence Minister Yuri Borisov told the Russian press that 'the Vitebsk suite will be modernized to work in [a] more broad band of frequencies and at long ranges, and to provide the best protection of aircraft against attacks of modern air-to-air and portable surface-to-air missiles'.

The most important element of the Monoblok suite is the new L418-5 directional infrared countermeasures module by SKB Zenit from Zelenograd; it is functionally similar to the previous L370-5 but is made on a new element base. The new turret has an angular shape, while the previous one was a spinning ball; inside, there is a new SP3-1500 lamp. The new L418-2 ultraviolet approach-warning sensors for the Monoblok system were made by the GIPO company in Kazan. The Aviaavtomatika company from Kursk is developing a new decoy launcher, SV370-418 (Sistema Vybrosa, launching system).

Individual components of the L418 Monoblok system, including L418-2E warning sensors and L418-5E jammers, were used on the Ka-52E helicopters delivered to Egypt. The system for the Russian MoD is still in trials, including on Ka-52K '104'. The first Ka-52M helicopters are equipped with the standard L370 Vitebsk devices.

Strelets-M C4ISTAR

Another new element in the modernised Ka-52M helicopter is its adaptation to work within the tactical battlefield air control system, Strelets-M (the similarity of the names with the 9S846 Strelets anti-aircraft missile launcher is accidental).

The KRUS Strelets (Kompleks Razvedki, Upravleniya i Svyazi; reconnaissance, control and communication system) has been in use since 2007, but for years it was only used by special forces. However, following the fighting in Ukraine from 2014 and in Syria from 2015, it was decided to implement it more widely among the ground troops; the currently used modernised version is the Strelets-M. For those who like abbreviations, the KRUS Strelets-M can be called the C4ISTAR, or Computers, Command, Control and Communications, Intelligence, Surveillance, Target Acquisition, and Reconnaissance system.

In the Strelets-M system, a soldier on the ground, after detecting a target with a portable radar and a laser rangefinder, introduces its coordinates on their own tablet. From there, they are transferred to the commander's portable computer and then transmitted to combat assets – artillery and aviation. Currently, in the Ka-52, the target coordinates are transmitted only by voice over the radio; in the Ka-52M, the target will be indicated automatically on the panel in the cockpit.

Chapter 10
What is next?

In August 2017, Russia's MoD commissioned the research work Skorost (speed) comprising the development of the concept of a new-generation combat helicopter. The main requirement of the new helicopter was a cruising speed of 249mph (400km/h). The Kamov company presented a project in its typical configuration with co-axial rotors, based on the current Ka-52; Mil developed the classic single-rotor concept. Common to both projects was the large wing to reduce the load on the rotor in high-speed flight. Note that on current Russian combat helicopters, the wing is primarily used for weapon carrying; the aerodynamic function is secondary.

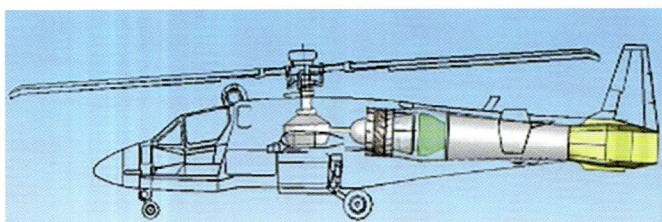

Left: An early design of a combined propulsion system for the Ka-52 with a jet nozzle. (Kamov)

Below: Planned conversion of Ka-52 into a technology demonstrator for a high-speed helicopter with a large aircraft-type wing and combined propulsion. (Kamov)

What is next?

While working on the concept of a high-speed combat helicopter, Kamov planned to build various experimental helicopters with combined propulsion using the Ka-52, but they all remained on paper. We learned of another project of Kamov from a patent registered in October 2016. It was a compound helicopter designed to reach a speed of 435mph (700km/h). The helicopter had co-axial main rotors and an airframe with a large wing, canards and double tailfins. The engines drove the main rotors and also provided thrust through the nozzles at the rear. In addition to the usual six underwing pylons, the helicopter had an internal weapon bay for eight LMUR missiles.

The results of Skorost research were approved at the end of 2018. Andrei Boginsky, the then general director of Russian Helicopters, announced the Mil concept had been selected and, in the next stage, the technical design and prototype of the helicopter were to be ordered. 'The choice was made primarily from the point of view of the feasibility of the project, the possibility of launching the helicopter production in 7–10 years,' he said. Three years have passed, and it is obvious this deadline will not be met and that the new-generation Russian combat helicopter will not be built soon.

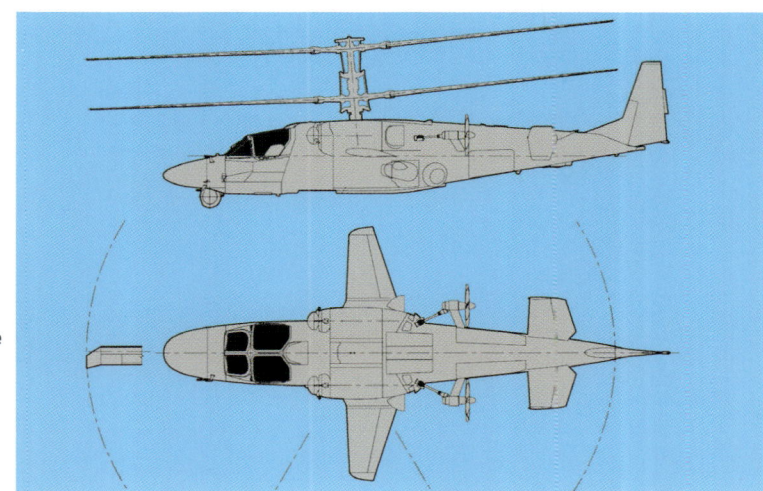

Right: Another way that has been considered to increase the speed of the Ka-52 is two additional pushing propellers on the fuselage sides. (Kamov)

Below: Patented by Kamov in 2016, the project of a combat helicopter reaching a speed of 700km/h. (Kamov)

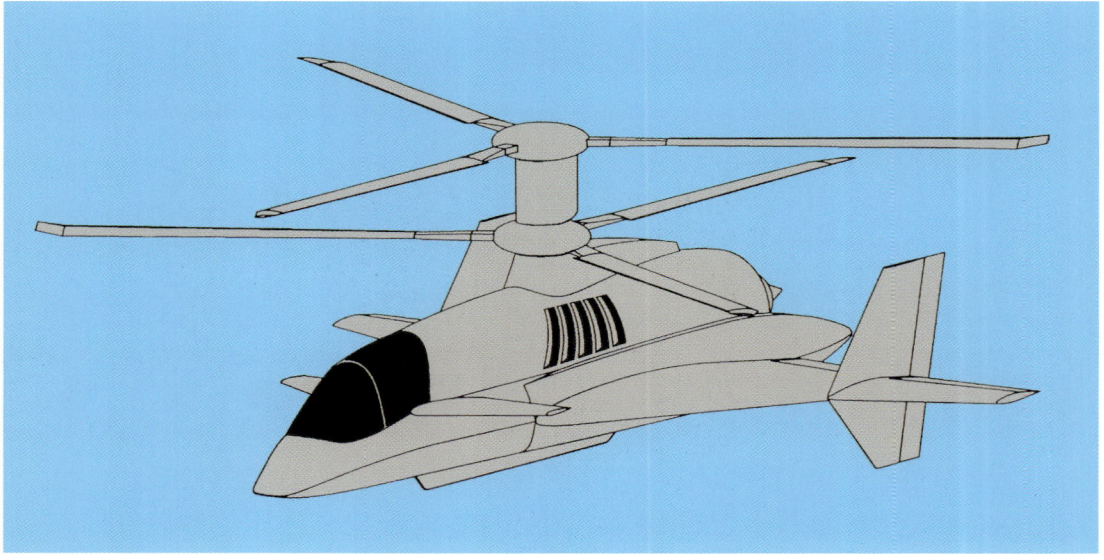

Other books you might like:

Modern Military Aircraft Series, Vol. 2

North Korean Aviation

Modern Military Aircraft Series, Vol. 1

Modern Military Aircraft Series, Vol. 4

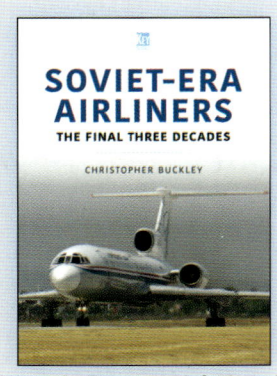

Historic Commercial Aircraft Series, Vol. 1

Airlines Series, Vol. 5

For our full range of titles please visit:
shop.keypublishing.com/books

VIP Book Club

Sign up today and receive
TWO FREE E-BOOKS

Be the first to find out about our forthcoming book releases and receive exclusive offers.

Register now at **keypublishing.com/vip-book-club**

Our VIP Book Club is a 100% spam-free zone, and we will never share your email with anyone else. You can read our full privacy policy at: privacy.keypublishing.com